Lecture Notes in Computer Science 4370

Commenced Publication in 1973
Founding and Former Series Editors:
Gerhard Goos, Juris Hartmanis, and Jan van Leeuwen

Pierre P Lévy   Bénédicte Le Grand
François Poulet   Michel Soto
Laszlo Darago   Laurent Toubiana
Jean-François Vibert (Eds.)

Pixelization Paradigm

First Visual Information Expert Workshop, VIEW 2006
Paris, France, April 24-25, 2006
Revised Selected Papers

 Springer

Volume Editors

Pierre P Lévy
AP-HP, Hôpital Tenon, INSERM UMR-S 707, 75970 Paris Cedex 20, France
E-mail: pierre.levy@tnn.aphp.fr

Bénédicte Le Grand
Michel Soto
Université Pierre et Marie Curie, LIP6-CNRS, 75005 Paris, France
E-mail: {benedicte.le-grand,michel.soto}@lip6.fr

François Poulet
ESIEA-Ouest, 53000 Laval, France
E-mail: poulet@esiea-ouest.fr

Laszlo Darago
Semmelweis University, 1082 Budapest, Hungary
E-mail: darago.laszlo@gmail.com

Laurent Toubiana
Université Paris 5, UPRES EA 222, 75743 Paris Cedex 15, France
E-mail: toubiana@necker.fr

Jean-François Vibert
AP-HP, Hôpital St Antoine, INSERM UMR-S 707, 75571 Paris Cedex 12, France
E-mail: jean-francois.vibert@upmc.fr

Library of Congress Control Number: 2007920819

CR Subject Classification (1998): I.4, I.5, I.2.10, I.3, I.2.6, F.2.2, J.3

LNCS Sublibrary: SL 6 – Image Processing, Computer Vision, Pattern Recognition, and Graphics

ISSN   0302-9743
ISBN-10   3-540-71026-4 Springer Berlin Heidelberg New York
ISBN-13   978-3-540-71026-4 Springer Berlin Heidelberg New York

Springer is a part of Springer Science+Business Media

springer.com

© Springer-Verlag Berlin Heidelberg 2007
Printed in Germany

Typesetting: Camera-ready by author, data conversion by Scientific Publishing Services, Chennai, India
Printed on acid-free paper   SPIN: 11945543   06/3142   5 4 3 2 1 0

Preface

The pixelization paradigm states as a postulate that pixelization methods are rich and are worth exploring as far as possible. In fact, we think that the strength of these methods lies in their simplicity, in their high-density way of information representation property and in their compatibility with neurocognitive processes.

- Simplicity, because pixelization belongs to two-dimensional information visualization methods and its main idea is identifying a "pixel" with an informational entity in order to translate a set of informational entities into an image.
- High-density way of information representation property, firstly because pixelization representation contains a third dimension—each pixel's color—and secondly because pixelization is a "compact" (two-dimensional) way of representing information compared with linear one-dimensional representations (Ganascia, p.255) .
- Compatibility with neurocognitive processes, firstly because we are three-dimensional beings and thus we are intrinsically better at grasping one- or two-dimensional data, and secondly because the cerebral cortex is typically a bi-dimensional structure where metaphorically the neurons can be assimilated to "pixels," whose activity plays the role of color (Lévy, p.3).

The pixelization paradigm may be studied along two related directions: *pixelization and its implementation* and *pixelization and cognition.*

The first direction—*pixelization and its implementation*—may be divided into two parts: *pixelization theory* and *pixelization application.*

Pixelization theory can itself be decomposed into three parts:

- Pixelization's mathematics (Lévy, p.3), which aims at formalizing and understanding the pixelization process. The potential fall-outs of this research axis are the creation of new automatic algorithms capable of building pregnant pixelized images and the application to neurocognitive processes understanding.
- Pixelization per se: this deals with the various methods, specific to pixelization, which improve its results. This concerns the grouping of data (Keim, p.12) or ordering of attributes (Abdullah, p.36), the reduction of large databases in order to pixelize and display them (Keim p.12, Poulet p.25), the association with statistical studies with spreadsheet software (Vidmar, p.50), the implementation in the context of an interactive temporal pixelized system (Gershkovich, p.57).
- Pixelization and multidimensional data: the relevance of this research direction lies in the high-density pixelization method property; and this is

clearly correlated with the problem of multidimensional data representation. The proposed approaches deal with the grouping of multidimensional data (Choong, p.65), the stacking of dimensions (Langton, p.79), the one-to-one mapping from a multidimensional space (Castro p.94) and the projection from this multidimensional space (Priam, p.108).

Pixelization applications can be decomposed into:

- Spatial pixelization where the support of the image has a spatial meaning. In the first paper the problem is to "build" the value of the pixel in a medical image fusion process (Montagner, p.121). The second paper proposes a pixel processing method to compare medical images (Ouchchane, p.136) and the third paper allows computing two-dimensional supports for shape matching of two molecules (Tripathi, p151).
- Temporal pixelization, where the support of the image has a temporal meaning. Various approaches are presented. The first paper proposes a "spiral" representation of time applied to course usage monitoring (Mazza, p.163), the second one proposes a linear representation of time applied to neural network activity displaying (Vibert, p.173) and the third one shows a bi-dimensional representation of time applied to the measurement of uterine electromyographic signal in sheep (Vidmar, p.183).
- Qualitative pixelization, where the support of the image has a purely qualitative meaning (i.e., neither temporal nor spatial). The first paper (Le Guillou, p.189) translates data and knowledge bases into pixelized images, the second paper (Jourdan, p.202) considers a two-dimensional scatter plot as an image and the last paper (Darago, p.217) proposes to identify medical information to a qualitative map.

The second direction corresponds to *neurocognitive processes*. The first paper uses pixelization as a tool to link high-level cognitive processes to low-level neurophysiological processes (Bernard p. 229, see also [1]). The second paper uses pixelization to visualize the activity of cortical layers (Abramov, p.242). The third paper (Ganascia, p.255) proposes a method for translating a text or a medium into a "color cognitive map." Finally the last paper (Trzaska, p.266) proposes a method allowing the evaluation of information visualization methods in general and pixelization methods in particular.

All these papers were the result of a rigorous reviewing process: all the papers were reviewed by at least three referees, 30 were accepted as oral communication and 23 were accepted for publication in the paper proceedings. Fifteen countries were represented.

Indeed this first workshop was a success and we wish to thank very sincerely all the members of the International Program Committee for their thorough review. It was a challenging approach and we think that new ways are now open. We are also very grateful to Springer for agreeing to publish these proceedings in their *Lecture Notes in Computer Science* series.

November 2006

Pierre P Lévy
Bénédicte Le Grand
François Poulet
Michel Soto
Laszlo Darago
Laurent Toubiana
Jean-François Vibert

[1] Lévy PP. Graduated Substrata. Information Processing & Management. Vol 24 n°6 pp 693-702, 1988.

Organization

Conference Chair

Pierre P Lévy, Assistance Publique Hôpitaux de Paris, INSERM, UPMC, France

Conference Co-chair

Laszlo Darago, University of Debrecen, Hungary

International Program Committee

Nadir Belkhiter	Université de Laval, Canada
Jean-Yves Boire	Université de Clermont-Ferrand, INSERM, France
Thanh Nghi Do	University of Can Tho, Vietnam
Gábor Fazekas	University of Debrecen, Hungary
Jean-Daniel Fekete	INRIA, Paris, France
Bernard Fertil	INSERM, Paris, France
Antoine Flahault	Assistance Publique Hôpitaux de Paris, INSERM, UPMC, France
Jean-Gabriel Ganascia	LIP6, Université Pierre et Marie Curie, Paris, France
Daniel Keim	University of Konstanz, Germany
Monique Noirhomme-Fraiture	Université de Namur, Belgium
Jean-Pierre Reveillès	Université de Clermont-Ferrand, France
Francis Roger France	Université Catholique de Louvain, Belgium
Christian Roux	ENST, Brest, France
Niilo Saranummi	VTT Information Technology, Finland
Jose Pedro Segundo	University of California, Los Angeles, USA
Simeon Simoff	University of Technology, Sydney, Australia
Annick Vignes	Université Paris II, UMR 7017, France
Allessandro Villa	Université Joseph Fourier, Grenoble 1, France

Local Organizing Committee

Bénédicte Le Grand	LIP6, Université Pierre et Marie Curie, Paris, France
François Poulet	ESIEA – Pôle ECD, Laval, France
Michel Soto	LIP6, Université René Descartes, Paris, France

Laurent Toubiana   INSERM U707, UPMC, Paris, France
Jean-François Vibert   Assistance Publique Hôpitaux de Paris, UPMC,
                                    INSERM, France

Sponsoring Institutions

Région Ile de France

ESIEA
Assistance Publique Hôpitaux de Paris
Hôpital Tenon
LIP6
INSERM
CNRS
IEEE EMBS

The Engineering in Medicine and Biology Society of the IEEE advances the application of engineering sciences and technology to medicine and biology, promotes the profession, and provides global leadership for the benefit of its members and humanity by disseminating knowledge, setting standards, fostering professional development, and recognizing excellence.

The field of interest of the IEEE Engineering in Medicine and Biology Society is the application of the concepts and methods of the physical and engineering sciences in biology and medicine. This covers a very broad spectrum ranging from formalized mathematical theory through experimental science and technological development to practical clinical applications. It includes support of scientific, technological, and educational activities.

Engineering in Medicine and Biology Society
IEEE
445 Hoes Lane
Piscataway, New Jersey, USA 08854
Telephone: +1 732 981 3433
Facsimile: +1 732 465 6435
E-mail: emb-exec@ieee.org

www.embs.org

Publications
Engineering in Medicine and Biology Magazine
IEEE Security & Privacy Magazine
Transactions on Biomedical Engineering
Transactions on Information Technology In Biomedicine
Transactions on Neural Systems and Rehabilitation Engineering
Transactions on Medical Imaging
Transactions on NanoBioscience
Transactions on Computational Biology and Bioinformatics

Electronic Products
EMBS Electronic Resource

Conferences
Annual International Conference of the IEEE Engineering in Medicine and Biology Society
IEEE EMBS Special Topic Conference on Microtechnologies in Medicine and Biology
IEEE EMBS Special Topic Conference on Neural Engineering
International Symposium on Biomedical Imaging (ISBI)
International Conference on Biomedical Robotics and Biomechatronics (BIOROB)
Bio, Micro and Nanosystems Conference

Summer Schools Sponsored by EMBS
International Summer School on Biomedical Imaging
International Summer School on Biomedical Signal Processing
International Summer School on Biocomplexity
International Summer School on Medical Devices and Biosensors
International Summer School on Applications of Information & Communication
Technology in Biomedicine

Table of Contents

1 Pixelization Theory

1.1 Pixelization

1.2 Pixelization and Multidimensional Data

2   Pixelization Applications

2.1   Spatial Pixelization

2.2   Temporal Pixelization

2.3   Qualitative Pixelization

3   Pixelization and Cognition

Pixelization Theory

Pixelization Paradigm: Outline of a Formal Approach

Pierre P. Lévy

Université Pierre et Marie Curie-Paris, UMR S 707, Paris F-75012 ;
INSERM, UMR-S 707, Paris, F75012 ;
Hôpital Tenon, Assistance Publique Hôpitaux de Paris, 4 rue de la Chine
75970 Paris cedex 20, France
pierre.levy@tnn.aphp.fr

Abstract. Various approaches exist related to Pixelization Paradigm. These approaches can be methodological or applied. Therefore finding formalism able to grasp all these approaches appears important. This formalism comes from category theory, the main formal tool being the functor notion. A category is characterized by objects and morphisms between these objects. Two categories can be associated thanks to a functor that respectively maps objects and morphisms from one category to objects and morphims of the other category. A pixelized visualization is defined by an object and two morphisms. The object is the reference frame of the visualization; the morphisms respectively associate a value and a colour to each cell of the reference frame. Functors formally allow implementing transformations on the visualization. The approach on the one hand allows identifying various atomic notions of Pixelization paradigm with formal entities and on the other hand, thanks to this, allows studying them and opening new ways. Various illustrations are given.

1 Introduction

We propose in this paper an outline of Pixelization Paradigm formalization. The aim of this approach is both theoretic and applied.

Theoretic by the way we believe in the heuristic value of theoretical approach: it is well known that generalizing a theory allows on the one hand enlarging the application field of this theory and on the other hand allows finding connections with other theories.

Applied, by the way this formalization, defining the formal atomic entities implied with Pixelization Paradigm, will constitute a facilitating computer programming frame. We will study accurately all these entities and this will open new ways that, we hope, will be explored during the next few years.

Our approach is to consider the set of papers proposed in the present workshop, through the proposed formal entities.

In what follows we will illustrate one by one these formal entities with notions developed in the papers. As often as possible we will try to show what problem has been solved by the author and what questions remain to be solved.

We will present firstly the formalism and secondly the results of applying this formalism to pixelization.

P.P. Lévy et al. (Eds.): VIEW 2006, LNCS 4370, pp. 3–11, 2007.

2 Methods: Formalism

The main idea of Pixelization Paradigm is to represent an informational entity by a pixel and this allows translating a set of information into an image. The strength of this approach is to create "a vivid semantic field" i.e. visualization. We describe below the mathematical entities generating this "vivid semantic field".

Definition 1
Let E be a set of informational entities and let S be a physical support. We call "infoxel" a couple (e, s) where e is an informational entity of E and s is an elementary cell of the support S to which e is identified.

This allows defining a reference frame R.

$$R = \{(e,s),(e,s) \in E \times S\}$$

In fact, R is the graph of bijection b that associates a cell s to each element e of E.

$$b : E \to S$$

$$e \mapsto s$$

An example is given in figure 1.

b	s1	s2	s3	s4
e1	x			
e2		x		
e3			x	
e4				x

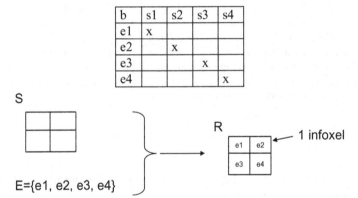

Fig. 1. Graphical representation of a reference frame R. Each informational entity of E is identified with a cell of the support S (they are labeled s1…s4 in the top table), providing an infoxel.

Definition 2
A Pixelized Visualization H is a triplet (R,μ,η) where:

- R is the reference frame of the Pixelized Visualization,
  $$R = \{(e,s),(e,s) \in E \times S\}$$
- μ is the function that associates each element (e,s) of the reference frame to a value d of D, a set of values.
  $$\mu : R \to D$$
  $$(e,s) \mapsto d_{es}$$
- η is a function from the set D of values to the set C of colors.
  $$\eta : D \to C$$
  $$d \mapsto c$$

Figure 2 continues the example given above.

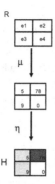

Fig. 2. Schematic representation of a pixelized visualization *H*

A category is characterized by objects and morphisms between these objects [7]. For example if we consider $\mathcal{S}$, Set Category, its objects are sets and its morphisms are functions defined on these sets. A set of axioms completes the definition of a category but they will not be used in the present work: our approach is only descriptive.

Two categories can be associated thanks to a functor that respectively maps objects and morphisms from one category to objects and morphims of the other category.

We can define a "Functor" from $\mathcal{S}$ to $\mathcal{S}$ as follows.

This functor F converts a pixelized visualization (R, μ, η) into another pixelized visualization (R',μ',η') in such a way that the diagram of figure 3 commute.

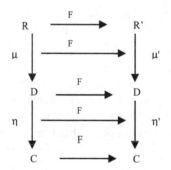

Fig. 3. Representation of functor F

Example

For example, a Functor F can be a formalization of "internal operations" on the Pixelized visualization [1]; these are the transformations [2] that are processed inside the image. In what follows we describe such a transformation where the informational entities are merged and where the associated values are summed.

By F, R is converted into R' as follows.

A set of infoxels $A=\{(s_i,e_i)\ i=1\ ...\ n\}$ is converted into a larger infoxel (S_A,E_A) where

$$S_A = \bigcup_{i=1}^{n}\{s_i\} \qquad \text{And} \qquad E_A = \bigcup_{i=1}^{n}\{e_i\}$$

This allows getting a reference frame R' where the granularity is coarser than the granularity in R.

An example continuing the above examples is given in figure 4.

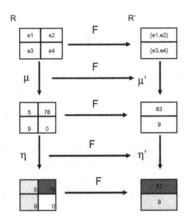

Fig. 4. Example of functor F. In this example, op_n is the sum (see equation (1) below).

By F the function μ is converted into μ' as follows

$$\mu'(F(A)) = F(\mu(A)) = F\left(\left\{d_{s_i e_i}, i = 1...n\right\}\right) = op_n\left(\left(d_{s_1 e_1}, d_{s_2 e_2},..., d_{s_n e_n}\right)\right) \qquad (1)$$

Where op_n is an n-ary operator (sum, mean or variance for example).

In the graphic of Figure 4 , $A=\{e1,e2\}$, $\mu(e_1)=5$, $\mu(e_2)=78$, $n=2$, $op_2=$sum

So $\mu'(F\{e_1,e_2\})= F(\mu(\{e_1,e_2\}))=F(\{5,78\})=sum((5,78))=83$

The function η is converted into η' in such a way as to be compatible with the new value range of D. In the above example, η' has been chosen to be equal to η.

3 Results

In what follows the above formalism is used to rigorously apprehend notions inherent to pixelization.

When we use pixelization, we have **a set of Entities E**: Entities can be various: from simple information to very complex structures. In particular, in the case of qualitative pixelization [p.189-226], the problem will be to define E. In other words to what informational entity e can a cell s of the physical support S be identified? Le guillou et al choose the triplet (case, feature, modality of the feature) [p.189].The quality of the visualization will depend on the relevance of the choice of E.

To visualize this set of entities, a **physical support S** is necessary.

This physical support S is made of "cells" which medium, size and shape may vary:

- Medium: paper, screen, cortical area in the brain. Paper is usually used for books, reports, journals. When the support is a screen, more possibilities are offered: clicking on the infoxels becomes possible to get more information; moreover a cinematographic view, i.e. using the temporal dimension, becomes possible. No paper was proposed about this in the present workshop, however this type of visualization is classically used, for example, for displaying epidemic spreading inside a geographic map [8]. The particular case of cortical area is worthwhile to be pointed: this emphasizes the connection of pixelization paradigm with neurophysiology. We will see it below.
- Size: a continuum from real pixel to "area".
  In some cases [p.12] all the infoxels will not have the same size and a semantic will be associated with this fact ("relevance" function of Keim, see below).
- Shape: square usually but all shapes can be seen (rectangle, hexagon, circle…).
  In some cases the shape and size of the infoxels can be different [p. 255].

3.1 The Bijection b

In some cases the relation b is not a bijective function; see the papers of Vidmar [p.50] and Darago [p.217] where each informational entity is associated to two cells. In this approach two infoxels correspond to each informational entity and two values are associated to each informational entity. A question would be to study the advantages and disadvantages of this approach.

An important point is the ordering of R. A useful ordering method is proposed by Abdullah and Hussain [p.36]. The Generalized Caseview method proposes a simple ordering method [4,5]. The quality of the visualization depends on the quality of the ordering.

In its own, this is a worthwhile way to be studied.

3.2 The Function μ

$$\mu : R \to D$$
$$(e,s) \mapsto d_{es}$$

In all visualizations we know, e is different from d_{se}: this emphasizes the very essence of pixelization methods: pixelization uses a physical support to visualize (present to the user) the datum that is associated to an informational entity. This is done by identifying the elementary physical entity of the physical support with the informational entity. Then each datum is materialized by a colour and this colour allows linking together the informational entities thanks to the spatial properties of the physical support. This is possible because some of the properties of the informational entities are converted into spatial bi-dimensional properties.

μ can be such that R is constant. In this case the approach is keeping the same reference frame and defining various μ corresponding to various visualization parameters: this is the case of ICD-view [p.217] or of the generalized Caseview method [5] or of spatial pixelization [p.121-162]. Note that in the case of the Generalized Caseview method, numerous μ will exist because this is the aim of the method to visualize various data using the same reference frame. The advantage of such an approach is that visualization becomes easier and easier because the reference frame remains the same until it is finally "built" in the mind of the user [6].

μ can be such that R is defined in each visualization. This is the case when visualizing various data bases with the same method, see for example the work of Mazza [p.163] where the visualized datum is always the same (number of access to a modules courses in a website), but where R, which depends on the number of considered courses and on the time range can change as a function of the user preferences. See also for example the work of Leguillou et al. [p.189] to process a data base, where the datum is the valuation given by an expert to a couple (feature, modality of the feature qualifying an object) or the work of Bernard et al. [p.229] for neurophysiological data, where the visualized parameter is the "energy" of the electroencephalographic signal at a given time and at a given frequency. In the case of multidimensional data a one to one mapping method allows associating an n-uple with a cell in the 2D support of the reference frame(Hilbert space filling curve, see the paper of Castro et al p.94). On the one hand the problem lies in the choice of the granularity of the 2D reference frame, which is correlated with the approximation level chosen when implementing the mapping. On the other hand the problem lies in the interpretation of this 2D reference frame by the user. A fruitful research field is open.

3.3 The Function η

$$\eta : D \to C$$
$$d \mapsto c$$

This function is very important: it is the "Third dimension" of the Pixelization paradigm. For a same set of information, it plays the role of a filter that allows showing relevant information according to the question to solve.

Usually η is very simple the user just having to define the colour as a function both of data and of his knowledge: however this allocation remains arbitrary.

On the other hand η can be more sophisticated using the HSB (Hue, Saturation and Brightness) encoding. For Mazza [p.163] the saturation corresponds to d and the user has to choose Hue and the scaling function allowing displaying the saturation optimally. This function can be linear or logarithmic.

Laurent et al. [p.65] use pixelization to visualize "blocks of data" inside a multidimensional space. The point is that they use HSB to encode both the intrinsic and the extrinsic properties of these blocks. The question to develop is the interpretation of the visualization result by a human; i.e. in other words what is the benefit of this sophisticated way of data representation for the user.

Thus when with Mazza colour meaning is intuitive and easy to grasp, with Laurent, the meaning, because very elaborated, is more difficult to grasp and needs learning. A promising way is open.

3.4 The Functor F

The notion of functor is important because on the one hand it corresponds to the notion of "internal operation" [1] on a pixelized visualization and on the other hand it represents a perceptual neuropsychological process.

In the data base visualization process it corresponds to reducing a large data base to a smaller one able to be displayed on a screen. In this case, Keim et al propose [p.12] a very sophisticated functor. This functor allows intelligent transformation of the initial R into a new F(R) where all infoxels have not the same size. This size both depends on a relevance function [3], which has been defined by the user, and on the grouping decisions implemented thanks to the logical tree that this relevance function allowed to build. In the example given by Keim $F(\mu)=op_v$ was the mean function, thus η had not to be modified by F. This type of functor is very interesting for very large data base visualization. As a matter of fact if the data base is too large (R,μ,η) is virtual and then, indeed, F is necessary to display the data.

A similar approach is adopted by Poulet [p.25] using symbolic data.

Another very sophisticated functor is proposed by Montagner et al.[p.121], in a context of spatial pixelization. Using our vocabulary, the problem can be stated as follows: given two images (R,μ) and (R',μ'), the resolution of R being higher than the resolution of R', is it possible to find a functor F allowing to transform (R,μ) into (R,μ'') where μ'' is a function resulting from the fusion of (R,μ) and (R',μ') (see Fig. 5).

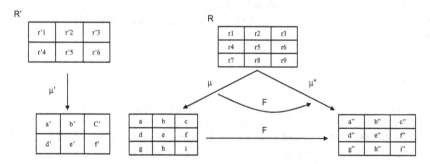

Fig. 5. Example of functor in spatial pixelization

Montagner et al. propose an explicit solution allowing to compute μ'' from μ and μ' on the one hand, and from the geometric relation between R and R' on the other hand (see equation (1) and Fig. 5 of their paper: $\mu=f$, $\mu'=g$ and $\mu''=f$).

This problem of image fusion is frequent in medicine and we think that the method could be applied to other fields.

Another example of functor appears in the work of Leguillou et al. [p.189] where a classification process might be described as a series of functors (see Fig. 11 of their paper).

Another example of functor comes from the work of Bernard et al. [p.229] where a series of pixelized visualization is proposed in fig.3; the passage from image 3b to image 3e corresponds to the application of functors. The interest of this approach is illustrating the various signal processing steps allowing the neurophysiologist to understand them. Indeed here a large field is open to pixelization methods.

4 Discussion – Conclusion

The proposed formalism has allowed us to give a panoramic view of pixelization problems.

Each time when pixelization method is applied the following questions ought to be asked. What is the problem to solve in view of the formalisms of the pixelization paradigm? Does the proposed solution bring a response to a general problematic of the pixelization paradigm? This is an important interest of this approach. As a matter of fact the proposed formalism allows locating each new work related to pixelization: if a new solution is proposed, it will be possible to spread this new solution to many related problems thanks to the formalism. For example, could the functor proposed by Montagnier et al. be applied to a problem of qualitative pixelization?

The present approach was only descriptive: the formal definition of a category [7] includes the definition of a composition law of the morphisms, of an axiom of associativity of this law and of an axiom of existence of the Identity morphism. Moreover the definition of a functor includes the preservation of the identity morphism and of the composition law. We were not interested in using all the formalism: this will be a next step; at our reasoning level we only gave a formal frame to pixelization paradigm. Exploring the effectiveness of the formal use of this tool will be the object of a future work.

References

[1] Bertin J. Semiology of Graphics: Diagrams, Networks, Maps (WJBerg, Trans.). Madison, WI: University of Wiskonsin Press (1967/1983)
[2] Card SK, Mackinlay JD and Shneiderman B. Information visualization. pp1-34. In : Readings in Information visualization, Using Vision to Think .Ed by S.K. Card, J.D Mackinlay and B. Shneiderman. San Diego, 1999.
[3] Keim DA, Kriegel H-P VisDB: Database Exploration using Multidimesnional Visualization. In : Readings in Information visualization, Using Vision to Think .Ed by S.K. Card, J.D Mackinlay and B. Shneiderman. San Diego, 1999.
[4] Lévy P.P The case view a generic method of visualization of the case mix. International journal of Medical Informatics 2004, 73, 713-718.
[5] Lévy P.P., Duché L., Darago L., Dorléans Y., Toubiana L., Vibert J-F, Flahault A.. ICP-Cview : visualizing the international Classification of Primary Care. Connecting Medical Informatics and Bio-Informatics. Proceedings of MIE2005. R. Engelbrecht et al. (Eds). IOS Press, 2005.pp623-628. 2005

[6] Lévy P.P. Caseview : building the reference set. In: Studies in health Technology and Informatics. Volume 105. Transformation of Healthcare with information technologies. IOS Press. Editors: Marius Duplaga, Krzysztof Zielinski, David Ingram.pp172-181, 2004.

[7] Whitehead GW Elements of Homotopy Theory. Graduate texts in Mathematics. Springer-Verlag. 1978.

[8] www.sentiweb.org

Scalable Pixel Based Visual Data Exploration

Daniel A. Keim, Jörn Schneidewind, and Mike Sips

University of Konstanz, Germany
{keim,schneide,sips}@dbvis.inf.uni-konstanz.de

Abstract. Pixel-based visualization techniques have proven to be of high value in visual data exploration, since mapping data points to pixels not only allows the analysis and visualization of large data sets, but also provides an intuitive way to convert raw data into a graphical form that often fosters new insights, encouraging the formation and validation of new hypotheses to the end of better problem solving and gaining deeper domain knowledge. But the ever increasing mass of information leads to new challenges on pixel-based techniques and concepts, since the volume, complexity and dynamic nature of today's scientific and commercial data sets are beyond the capability of many of current presentation techniques. Most existing pixel based approaches do not scale well on such large data sets as visual representation suffers from the high number of relevant data points, that might be even higher than the available monitor resolution and does therefore not allow a direct mapping of all data points to pixels on the display. In this paper we focuses on ways to increase the scalability of pixel based approaches by integrating relevance driven techniques into the visualization process. We provide first examples for effective scalable pixel based visualizations of financial- and geo-spatial data.

1 Introduction

Due to the progress in computer power and storage capacity over the last decade, today's scientific and commercial applications are capable of generating, storing and processing massive amounts of data. Examples are historical data sets including census data, financial data or transaction data from credit card-, telephone- and e-commerce companies. Additionally there exist many dynamic processes, arising in business, network or telecommunication, which generate tremendous streams of time related or real time data like sensor data, web click streams or network traffic logs.

The analysis of such massive data sets is an important and challenging task, since researchers and analysts are interested in patterns in the data, including associations, correlations or exceptions. These information is needed in order to turn the collected data into knowledge, e.g. to identify bottlenecks, critical process states, fraud or any other interesting information hidden in the data.

Visualization techniques have been proven to be of great value in supporting the data exploration process, since presenting data in an interactive, graphical form often fosters new insights, encouraging the formation and validation of new hypotheses to the end of better problem solving and gaining deeper domain knowledge [1].

P.P. Lévy et al. (Eds.): VIEW 2006, LNCS 4370, pp. 12–24, 2007.
© Springer-Verlag Berlin Heidelberg 2007

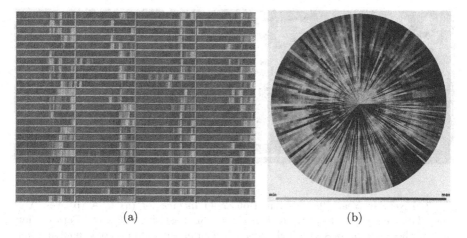

(a)   (b)

Fig. 1. (a) Dense Pixel Displays: Recursive Pattern Technique showing 50 stocks in the FAZ (Frankfurt Stock Index Jan 1975 - April 1995). The technique maps each stock value to a colored pixel; high values correspond to bright colors.©IEEE (b) CircleView showing stock prices of 240 stocks from the S&P 500 over 6 months.

1.1   Pixel Based Visualization Techniques

Pixel-oriented techniques are a special group of visual data mining techniques and are especially important for visualizing very large multidimensional data sets. The general idea of pixel oriented visualization techniques is to represent as many data objects as possible on the screen at the same time by mapping each data value to a colored pixel on the screen and arranging the pixels adequately, e.g. by grouping the pixels belonging to each dimension into adjacent areas. Therefore the techniques allow the visualization of the maximum amount of data on current displays , up to about 1.000.000 data values at typical monitor resolution [2]. Many pixel-oriented visualization techniques have been proposed in recent years and it has been shown that the techniques are useful for visual exploration of large databases in a number of different applications [3,4] (see CircleSegments, Recursive Pattern or the CircleView technique [1,5]). The *Recursive Pattern technique* for example is based on a generic recursive back-and-forth arrangement of the pixels and is particularly aimed at representing datasets with a natural order according to one attribute (e.g. time-series data). In Figure 1(a), an recursive pattern visualization of financial data is shown. The visualization shows twenty years (January 1974 - April 1995) of daily prices of the 50 stocks contained in the Frankfurt Stock Index (FAZ). All pixel-display techniques partition the screen into multiple subwindows. For data sets with m dimensions (attribute), the screen is partitioned into m subwindows, one for each of the dimensions. Correlations, functional dependencies, and other interesting relationships between dimensions may be detected by relating corresponding regions in the multiple windows. To achieve that objective, a number of design problems have to be solved. These design issues include color mapping,

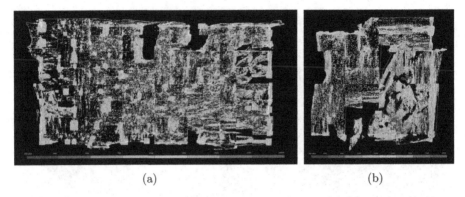

(a)   (b)

Fig. 2. PixelMap Year 1999 Median Household Income for (a) USA (b) New York State. This map displays cluster regions e.g. on the East side of Central Park in Manhattan, where inhabitants with high income live, or on the right side of Brooklyn, where inhabitants with low income live.

arrangements of pixels, the shape of the subwindows and the ordering of dimensions. An overview on pixel-oriented visualization techniques and more details about designing pixel based visualization methods can be found in [1].

1.2   Spatial Pixel Based Techniques

In many application domains, data is collected and referenced by geo-spatial locations. Spatial data mining, or the discovery of interesting patterns in such databases, is an important capability in the development of database systems. The problem of visualizing geo-referenced data can be described as a mapping of input data points, with their associated original positions and statistical attributes, to unique positions on the output map. Consider, for example credit card purchase transactions including both, the address of the place of purchase and of the purchaser; telephone records including caller addresses or cell phone base antennae locations; space satellite remote sensed data; census and other government statistics with addresses or other geo-graphic indexes for residents; or records of property ownership based on physical locations. Often, discovering spatial patterns is crucial for understanding these data sets. Pixel based spatial techniques like the PixelMap [6] approach, a way of displaying dense point sets on maps which combines clustering and visualization, do not aggregate the data, instead they map the data points directly on the output map, by considering constraints like no overlap, position preservation and clustering of data points. Figure 2(a) shows a PixelMap visualization of the Year 1999 Median Household Income Data for the US and 2(b) for the US-State of New York. Pixels represent households and color shows the household income. It is easy to see that households with very high median household income are located in Manhattan and Queens, and households with low median household income are in the Bronx and Brooklyn. Especially, very wealthy inhabitants live on the east side of Central Park.

1.3 Challenges on Pixel Based Techniques

The successful application of dense pixel displays and pixel based geo-spatial techniques in various visual analysis scenarios has shown that pixel techniques are a powerful way to explore large data sets. However, the increasing size and complexity of today's data sets leads to new challenges for pixel based techniques in coping with scale. Therefore Visual Analytics, which aims on the integration of data mining technology and information visualization to deal with the rapidly growing amounts of data in very different areas, is becomig an increasingly important topic. Visual Analytics will incorporate more "intelligent" means than to just retrieve and display a set of data items [7]. Automatic analysis techniques have to be tightly integrated with advanced visualization and interaction techniques. In this context the scalability of visualization techniques in problem size is among the top 10 Visual Analytics Research Challenges[8], in order to keep step with the growing flood of information.

Eick and Karr[9] proposed an scalability analysis and came to the conclusion that many visualization metaphors do not scale effectively, even for moderately sized data sets. Scatterplots for example, one of the most useful graphical techniques for understanding relationships between two variables, can be overwhelmed by a few thousand points. Additionally, there are two limiting factors for all pixel based visualization techniques: human perception and display area. On one hand, human perception, that means the precision of the eye and the ability of the human mind to process visual patterns, limits the number of perceptible pixels and therefore affects visual scalability directly. One the other hand, monitor resolution affects on visual scalability through both physical size of displays and pixel resolution. At a normal monitor viewing distance, calculations in [10] suggest that approximately 6.5 million pixels might be perceivable for the human, given sufficient monitor resolution [9]. Since most of today's visualization tools have been optimized for conventional desktop displays with ca. 2 million pixels, in typical application scenarios monitor resolution rather than human vision is the limiting factor.

Based on these facts, the analysis of large data sets reveals two major tasks. The first one is the question, how visualizations for massive data sets can be constructed without loosing important information, even if the number of data points is to large to visualize each single data point at full detail or without overlap. The second important task is to find techniques to efficiently navigate and query such large data sets.

2 Scalability Issues

In the context of Pixel based techniques, we define scalability as the ease with which a pixel based visualization technique or metaphor can be modified to fit the problem area. In our experiments we observed three major issues concerning the scalability of pixel based techniques, briefly described in the following. Based on these observations, we introduce in Section 3 the concept of relevance driven pixel visualizations and present first experimental results in Section 4.

Fig. 3. Large Scale Displays provide new possibilities in visual data exploration, e.g. the visualization of millions of data points and finer structures

2.1 High Resolution Displays

A straight forward, but not always practicable solution to increase the number of pixels that can be directly placed on the display is the use of high resolution displays. Wall-sized pixilated displays provide a powerful opportunity to visualize large data sets. The resolution of these displays is usually up to 10 times higher than on normal desktop displays which opens new possibilities in visual data exploration. The basic idea is to provide more pixels to represent more data points. That means that dense pixel displays may place a much larger number of data points in a single view and scalability can be increased directly. Figure 3 shows the *iWall* at the University of Konstanz [11] representing a PixelMap visualization. To point out the difference in size, in the foreground the same results is visualized on a normal desktop monitor. For geo-spatial pixel techniques, where pixels may overlap because they refer to the same geo-location, higher display resolution does not necessarily mean a lower pixel overlap. Thus, while extra display pixels allow to show more data points, this technology alone does not eliminate occlusion. To solve this overlap problems by taking constraints like clustering or position preservation into account, algorithms are necessary that are still computational expensive even on high resolution visualizations.

2.2 Large Data Volumes

The more adaptive approach to apply pixel based techniques to very large data sets, where not every data point can be directly represented as pixel on the display, is to reduce the data volume to an appropriate size. This can be done in numerous ways, e.g. by employing Sampling techniques, Selection approaches, Multiresolution approaches like Wavelet Analysis or Hierarchical Aggregation techniques and a number of visualization techniques that focus this task have been proposed [12,13]. Although these techniques may efficiently reduce the data volume, there is always the risk of loosing important information, i.e. during the selection, transformation or aggregation process.

2.3 Multiple Dimensions

Another problem in analysing complex data sets is that the data is often high dimensional, i.e. for each data point there are a large number of attributes representing the object. In today's data sets the number of dimensions reaches hundreds or even thousands, and their analysis and visualization can become difficult. Irrelevant features may also reduce the accuracy of some algorithms. To address this problem, a common approach is to identify the most important features associated with an object so that further processing can be simplified without compromising the quality of the final results. Besides manual attribute selections based on existing domain knowledge, there exist several way's for automated dimension reduction, like the well-known PCA. A technique for Visual hierarchical dimension reduction is proposed in [14].

3 Relevance Based Pixel Visualization

To increase scalability of existing pixel based techniques, the paper focuses on providing relevance driven compact representations of the data, which can than directly be mapped to the display space. The basic idea of relevance based pixel visualization is to decompose the data display space into local screen regions with individual object resolutions. These object resolutions control the granularity of the data points within each particular region. To provide and manage the different data granularities, a tree structure is employed. The structure of the tree highly depends on predefined analytical objective functions, which determine the relevance of single- or sets of datapoints. The goal is to provide an initial visual presentation of the whole data set with respect to the available screen space , that gives relevant parts of the data more space on the screen to present them at higher detail. As an example, we consider a clustering method, which determines clusters in a given data set as shown in Figure 4. Without loss of generality we assume that the data set contains 2-dimensional data. A simple way to visually detect clusters in the data are classical scatterplots. But if the number of data points is high, overlap in some portions of the display would it make really hard to detect clusters or to distinguish clusters from noise. Our approach instead determines a relevance value for each data point, e.g. the number of data points of the corresponding cluster. Based on the relevance values of the data points, a hierarchy is created, similar to single linkage clustering. In each step the data subsets with the lowest average importance value are merged,

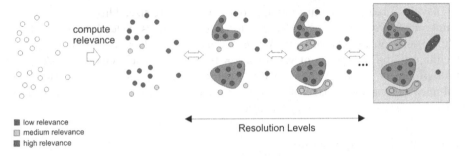

low relevance
medium relevance
high relevance

Resolution Levels

Fig. 4. *Basic idea* – decomposition into local screen regions allows it to display an overview of the entire data set

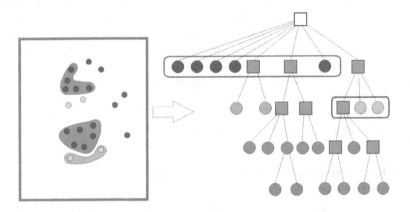

Fig. 5. *Hierarchy of MultiResolution Objects (MRO)* – The relevance value of the MRO's decreases with increasing tree level, the nodes in the grey boxes are selected for visualization

resulting in a hierarchical structure with different levels of detail like shown in Figure 5. To visualize the data set, data objects from the tree structure are selected so that the number and relevance of the selected objects is maximized, depending on the given display space. Figure 5 indicates a possible selection. The construction of the hierarchy is independent from the underlying visualization, and may therefore be combined with most visualization techniques to increase scalability.

3.1 Basic Concepts

Like described in the last section our approach consists of 3 steps: 1. Determine the relevance of each data point in the data set based on a given relevance function 2. Provide a compact hierarchical representation of the data with different levels of detail based on the relevance of data points 3. Select a given number of objects from the hierarchy so that the relevance is maximized and each data point is contained in at least one selected data object. The relevance function allows us to determine the object resolution of the local screen spaces. Of course the relevance function depends on the application scenario and must be determined by the user. Further research will focus on providing predefined relevance function for certain tasks like clustering or outlier detection. In general the relevance function is defined as:

Definition 1. *Relevance Function*
Let $A = \{a_0, \cdots, a_{N-1}\}$. The relevance function $\psi : A \longrightarrow \mathbf{R}$ assigns every data point $a_i \in A$ a relevance value $\psi(a_i)$.

Based on the relevance of single data points we are now able to construct an hierarchical structure by merging or combining data points with lowest relevance values in order to get a compact representation of the data set. Therefore the Multi-Resolution Objects are defined as:

Definition 2. *Multi-Resolution Objects*
Let $A = \{a_0, \cdots, a_{N-1}\}$ be the input data points and $\Psi = \{\psi(a_0), \cdots, \psi(a_{N-1})\}$ their associated relevance values. A MultiResolution object MRO is a set of locally close data points which have similar relevance values

$$MRO = \{a_i \in A : \forall a_j \in MRO : |\psi(a_i) - \psi(a_j)| \leq \kappa \wedge d(a_i, a_j) \leq \epsilon\}$$

Within every multiresolution object we define a object resolution level l_i which is application dependent. We suggest to identify the object resolution level l_i as the average of the relevance of all multi-resolution object members. Application dependent other functions (e.g. *min* or *max*) may be used.

Definition 3. *Object Resolution*
Let $mro_h = \{a_i, \cdots, a_j\}$ be a MRO and $\Psi = \{\psi(a_i), \cdots, \psi(a_j)\}$ the associated relevance values of the members of the multi-resolution object mro_i. The object resolution level l_i can be determined as:

$$l_h = \frac{\sum_{k=i}^{j} \psi(a_k)}{N}$$

For a given number of desired objects n_{desire}, the MultiResolution object tree is then constructed similar to Agglomerative Clustering: Each object of the data set is initially placed into its own MRO. Because the dataset A is assumed to be massive, the number of pixels is not sufficient to place all data points (Single MRO's) directly to the screen space: $|DS| << |A|$. Therefore we iteratively merge closest pairs of objects with low relevance values into a single group and create a pointer to the original objects. The iteration stops if the data set is reduced to the number of desired data points. These MultiResolution objects can than be mapped to teh display space. Considering that $MRO = \{mro_1, \cdots, mro_m\}$ is a decomposition of the data space into a set of multi-resolution objects, the goal of the multi-resolution visualization is to determine an useful mapping function f of the MRO objects to the display space DS that must satisfy four visual exploration goals: Relevance Preservation, Minimal Decomposition, Space Filling and No Overlap. In the next section we provide 2 examples where we adapted the introduced concepts to real world application scenarios, resulting in pixel techniques that scale well according to the described scalability issues.

4   Application Examples

4.1   MultiResolution CircleView

We integrated the Multiresolution approach into the CircleView technique in order to analyse a historical stock market dataset containing S&P 500 stock prices. The basic idea of the CircleView visualization technique is to display object attributes as segments of a circle. If the data consists of k dimensional attributes, the circle is partitioned into k segments, each representing the distances for one attribute. Inside the segments, the distance values belonging to one attribute are arranged from the center of the circle to the outside in a subsegment layout. The size of the segments and subsegments can either be predefined or parameter dependent. Additionally the size of each subsegment can vary from pixel to segment size. CircleView supports ordering and clustering of segments and subsegments, user interaction as well as nearest neighbour searches between segments or within single segments.

Figure 6 presents the basic idea of MultiResolution CircleView, showing the stock prices of 240 stocks from the S&P 500 over 6 months, starting from January 2004 (outside of the circle) to June 2004 (center of the circle). Each segment represents a single stock and each subsegment the average closing stock price of a certain time period depending on the detail level. The number of data values that can be visualized without aggregation using CircleView is limited by the circle area $C_{area} = 2 * \pi * radius$. In massive datasets this border can easily be reached. That means, if the radius of our circle would be less than 120 pixels, it is impossible to show all 120 stock prices without occlusion. Since from an analysts point of view it may be more interesting to analyse actual stock prizes rather than historic ones, the basic idea is to show only actual stock prizes at full detail and to present historic values as aggregated high level views. That is, the relevance value of each data point per stock (stock price) is determined by it's time stamp.

In Figure 6 the five latest day closing stock prices are shown at full detail by the innermost subsegments. The level of detail as well as the length of the subsegments decrease from the center to the outside of the circle. Historic stock prices are only presented as average values per week, per month or per year depending on the particular data as shown in Figure 6. The user is able to access particular information on actual data instantly, e.g. by mouse interaction and

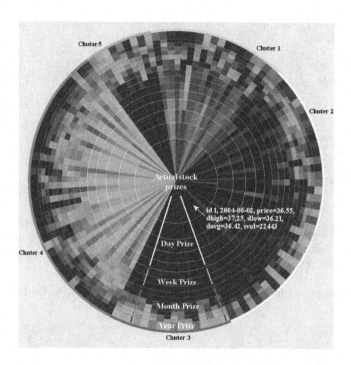

Fig. 6. CircleView showing stock prices of 240 stocks from the S&P 500 over 6 months (120 business days). Each Segment represents a single stock, the subsegments represent stock prices. The most actual stock price are shown in the middle of the circle at full detail. For older stock prices the multiresolution approach presents average prices per week/month/year. Stocks are clustered, Color indicates stock value.

gets at the same time an overview of the whole dataset. Drill down operations on items with lower resolution provide detailed information on historic data on demand.

4.2 Scalable PixelMaps

In Figure 7 a PixelMap visualization of the Year 2000 U.S. Median Household Income data is shown. Depending on the relevance value of every individual data point the resulting PixelMap shows the median household income of every individual block or the aggregated income of some map regions. The relevance value depends on the area of interest and in the upper we are interested to see details about the east coast. Figure 8 shows PixelMaps with different degrees of overlap and distortion. The data analyst can interactively chose the appropriate distortion level to control the fine structures on the screen. Note, to compute a pixel-layout without any overlap can be very time consuming, and sometimes it is not necessary to see such fine structures. For example, a quick overview about the data computed in real-time to identify interesting map regions (all regions have the same relevance value).

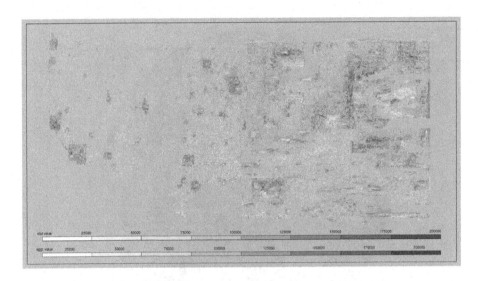

Fig. 7. Interactive PixelMaps – focus on household income of the US-East Coast

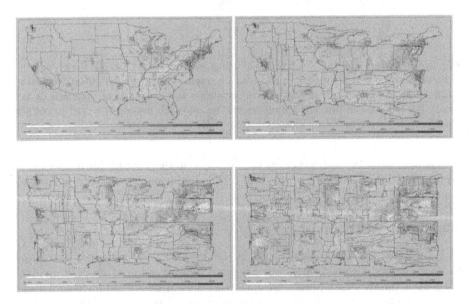

Fig. 8. Scalable PixelMaps – different degrees of overlap and distortion. Individual data points are going to be aggregated if the number of data points is greater than the available screen space to provide interactive displays.

We can easily see that New York City and Los Angeles County are the population areas of greatest interest in the United States. We can even see the distribution of U.S. Year 2000 Median Household Incomes in these regions. Finally, for the U.S. Year 2000Median Household Income we can observe a sharp decline between high and low income.Most Americans had income below \$75000 U.S., and ca. 10% of all Americans live below the federal poverty level.

5   Conclusion

The paper focuses on scalability issues of existing pixel based techniques in the context of visual exploration of large data sets. Since many of today's visual exploration techniques do not scale well on large data sets, we provide a hierarchical technique for reducing data size, by relevance driven data accumulation. We integrated this techniques into the CircleView and PixelMap approaches and applied them to a financial and a geo-spatial data set.

References

1. Daniel A. Keim. Designing pixel-oriented visualization techniques: Theory and applications. *IEEE Transactions on Visualization and Computer Graphics*, 6(1):59–78, 2000.
2. Jean-Daniel Fekete and Catherine Plaisant. Interactive information visualization of a million items. In *INFOVIS '02: Proceedings of the IEEE Symposium on Information Visualization (InfoVis'02)*, page 117, Washington, DC, USA, 2002. IEEE Computer Society.
3. Colin Ware. *Information visualization: perception for design*. Morgan Kaufmann Publishers Inc., San Francisco, CA, USA, 2000.
4. Stuart K. Card, Jock D. Mackinlay, and Ben Shneiderman, editors. *Readings in information visualization: using vision to think*. Morgan Kaufmann Publishers Inc., San Francisco, CA, USA, 1999.
5. D. A. Keim, J. Schneidewind, and M. Sips. CircleView: a new approach for visualizing time-related multidimensional data sets. In *Proceedings of the working conference on Advanced visual interfaces AVI, Gallipoli, Italy*, pages 179–182, 2004.
6. Daniel A. Keim, Christian Panse, Mike Sips, and Stephen C. North. Pixelmaps: A new visual data mining approach for analyzing large spatial data sets. In *ICDM '03: Proceedings of the Third IEEE International Conference on Data Mining*, page 565, Washington, DC, USA, 2003. IEEE Computer Society.
7. J. J. Thomas. Visual Analytics: a grand challenge in science - turning information overload into the opportunity of the decade. In *Keynote at IEEE Symposium on Information Visualization, Minneapolis, MN*, 2005.
8. Daniel A. Keim. Scaling visual analytics to very large data sets. *Workshop on Visual Analytics, Darmstadt, Germany*, 2005.
9. S. Eick and A. Karr. Visual scalability. *In J. of Computational and Graphical Statistics*, 1(11):22–43, 2002.
10. E. Wegman. Huge data sets and the frontiers of computational feasibility, 1995.
11. Databases, Data Mining and Visualization Research Group. Databases, Data Mining and Visualization Research Group iWall website, 2005. http://dbvis.inf.uni-konstanz.de/, Jul. 2005.
12. Pak Chung Wong, A. H. Crabb, and R. D. Bergeron. Dual multiresolution hyperslice for multivariate data visualization. In *INFOVIS '96: Proceedings of the 1996 IEEE Symposium on Information Visualization (INFOVIS '96)*, page 74, Washington, DC, USA, 1996. IEEE Computer Society.

13. D. Jerding and J. Stasko. The information mural: A technique for displaying and navigating large information spaces. *IEEE Transactions on Visualization and Computer Graphics, 4(3)*, 1998.

14. J. Yang, M. O. Ward, E. A. Rundensteiner, and S. Huang. Visual hierarchical dimension reduction for exploration of high dimensional datasets. In *VISSYM '03: Proceedings of the symposium on Data visualisation 2003*, pages 19–28. Eurographics Association, 2003.

High Dimensional Visual Data Classification

François Poulet

ESIEA
38, rue des Docteurs Calmette et Guérin
Parc Universitaire de Laval-Changé
53000 Laval – France
poulet@esiea-ouest.fr

Abstract. We present new visual data mining algorithms for interactive decision tree construction with large datasets. The size of data stored in the world is constantly increasing but the limits of current visual data mining (and visualization) methods concerning the number of items and dimensions of the dataset treated are well known (even with pixellisation methods). One solution to improve these methods is to use a higher-level representation of the data, for example a symbolic data representation. Our new interactive decision tree construction algorithms deal with interval and taxonomical data. With such a representation, we are able to deal with potentially very large datasets because we do not use the original data but higher-level data representation. Interactive algorithms are examples of new data mining approach aiming at involving more intensively the user in the process. The main advantages of this user-centered approach are the increased confidence and comprehensibility of the obtained model, because the user was involved in its construction and the possible use of human pattern recognition capabilities. We present some results we obtained on very large datasets.

1 Introduction

The size of data stored in the world is constantly increasing (data volume doubles every 9 months world-wide) but data do not become useful until some of the information they carry is extracted. Furthermore, a page of information is easy to explore, but when the information reaches the size of a book, or library, or even larger, it may be difficult to find known items or to get an overview. Knowledge Discovery in Databases (KDD) can be defined as "the non-trivial process of identifying valid, novel, potentially useful, and ultimately understandable patterns in data" [Fayyad et al, 1996].

In this process, data mining can be defined as the particular pattern recognition task. It uses different algorithms for classification, regression, clustering or association. In usual KDD approaches, visualization tools are only used in two particular steps:

- in one of the first steps to visualize the data or data distribution,
- in one of the last steps to visualize the results of the data mining algorithm,
  between these two steps, automatic data mining algorithms are carried out.

P.P. Lévy et al. (Eds.): VIEW 2006, LNCS 4370, pp. 25–34, 2007.

Some new methods have recently appeared [Wong, 1999], [Ankerst et al, 2001], [Poulet, 2004], trying to involve more significantly the user in the data mining process and using more intensively the visualization [Aggarwal, 2001], [Shneiderman, 2002], this new kind of approach is called visual data mining. We present a graphical method we have developed to increase the visualization part in the data mining process and more precisely in supervised classification tasks.

This method is an extension of a previously existing one PBC (Perception Based Classification) [Ankerst, 2000] (itself derived from the Circle Segments [Keim et al, 1995]) with the use of symbolic data [Bock and Diday, 2000] and more particularly interval and taxonomical data. The pixellisation techniques used in the Circle Segments and PBC are known to allow to deal with larger data sets than other interactive data mining methods. But even with these methods a dataset such as Forest Cover Type (581.000 instances, 54 attributes, 7 classes) from the UCI machine learning repository [Blake and Merz, 1998] requires about 20 screens to be displayed.

To overcome this drawback, we use a higher-level data representation, we do not deal with the original data but with a new data set created with symbolic data (for example, interval-valued data). The size of the dataset can be significantly reduced without loosing information and thus we can deal with very large number of items in the datasets.

In section 2 we present PBC, an interactive decision tree construction algorithm then in section 3 we present interval-valued and taxonomical data. The section 4 gives some details about the new interactive decision tree algorithms with interval-valued and taxonomical data and some results we have obtained before the conclusion and future work.

2   Interactive Decision Tree Construction

New interactive decision tree construction algorithms have appeared recently: PBC (Perception Based Classification) [Ankerst, 2000], DTViz (Decision Tree Visualization) [Han, Cercone, 2001], [Ware et al, 2001] and CIAD [Poulet, 2001]. All these algorithms try to involve more significantly the user in the data model construction process.

The technical part of these algorithms are somewhat different: PBC and DTViz use an univariate decision tree by choosing split points on numeric attributes in an interactive visualization. They use a bar visualization of the data: within a bar, the attribute values are sorted and mapped to pixels in a line-by-line (or pixel-by-pixel) fashion according to their order. Each attribute is visualized in an independent bar. As shown in figure 1, the first step is to sort the pairs (attr$_i$, class) according to attribute values, and then to map to vertical lines colored according to class values. When the data set number of items is too large, each pair (attr$_i$, class) of the data set is represented with a pixel instead of a line. Once all the bars have been created (fig.2), the interactive algorithm can start. The classification algorithm performs univariate splits and allows binary splits as well as n-ary splits (fig.3).

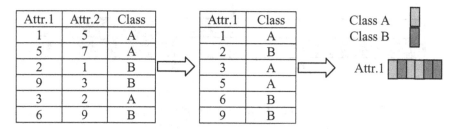

Attr.1	Attr.2	Class
1	5	A
5	7	A
2	1	B
9	3	B
3	2	A
6	9	B

Attr.1	Class
1	A
2	B
3	A
5	A
6	B
9	B

Class A
Class B

Attr.1

Fig. 1. Creation of the bar chart in PBC

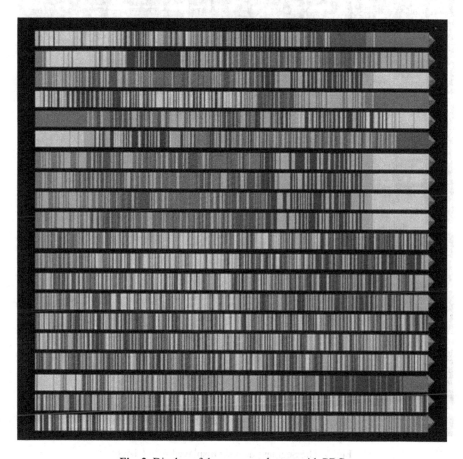

Fig. 2. Display of the segment dataset with PBC

CIAD is a bivariate decision tree using line drawing in a set of two-dimensional matrices: the scatter plot matrices [Carr et al. 1987]. The first step of the algorithm is the creation of a set of $0.5*(n-1)^2$ two-dimensional matrices (n being the number of attributes). These matrices are the two dimensional projections of all possible pairs of attributes, the color of the point corresponds to the class value.

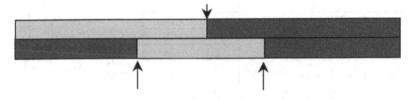

Fig. 3. PBC possible splits

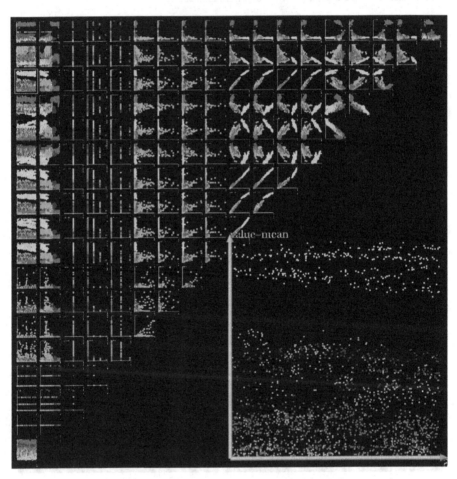

Fig. 4. The Segment data set displayed with CIAD

This is a very effective way to graphically discover relationships between two quantitative attributes. One particular matrix can be selected and displayed in a larger size in the bottom right part of the view (as shown in figure 4 using the Segment data set from the UCI repository [Blake and Mertz, 1998], it is made of 19 continuous attributes, 7 classes and 2310 instances).

Then the user can start the interactive decision tree construction by drawing a line in the selected matrix and performing thus a binary, univariate or bivariate split in the current node of the tree.

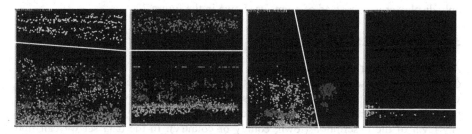

Fig. 5. The first splits performed with CIAD on the Segment dataset

Only PBC and CIAD provide the user with an automatic algorithm to help him choose the best split in a given tree node. The other algorithms can only be run in a 100% manual interactive way.

3 Symbolic Data

3.1 Interval Data

Decision trees usually deal with qualitative or quantitative values. Here we are interested in interval-valued data. We only consider the particular case of finite intervals. Compared to existing automatic decision tree algorithms, our approach is similar to the one of [Ciampi et al, 2000] and [Asseraf et al, 2004]. In the case of CIAD extension, the cut is performed for a given attribute value like in [Ciampi et al, 2000] and in the case of PBC extension the cut is performed according to a given item (interval-valued) like in [Asseraf et al, 2004].

To be able to use this new kind of data with PBC, we need to define an order on these data. There are mainly three different orders we can use [Mballo and Diday, 2004]: according to the minimum values, the maximum values or the mean values.

Let us consider two interval data: $i_1=[l_1,r_1]$ (mean=m_1) and $i_2=[l_2,r_2]$ (mean=m_2).
If the data are sorted according to the minimum values, then:
if $l_1= l_2$, then $i_1 < i_2 <=> r_1< r_2$; if $l_1 \neq l_2$, then $i_1 < i_2 <=> l_1< l_2$.

If the data are sorted according to the maximum values, then:
if $r_1= r_2$, then $i_1 < i_2 <=> l_1< l_2$; if $r_1 \neq r_2$, then $i_1 < i_2 <=> r_1< r_2$.

And finally, if the data are sorted according to the mean values, then $i_1<i_2 <=> m_1<m_2$.

We can choose any of these three functions to create the bar in the first step of the PBC algorithm in order to sort the data according to the values of the current interval-valued attribute.

This method is used in the first step of the PBC algorithm to create the bar charts. Once this task has been performed for each attribute, the classification algorithm is exactly the same as for continuous data (when it is used in its 100% manual mode). Concerning CIAD, the tree is created exactly the same way as for continuous data, we only display crosses instead of points as shown in the figure 7.

3.2 Taxonomical Data

A taxonomical variable can be defined [Bock, Diday, 2000] as a mapping of the original data on a set of ordered values. It is equivalent to a structured or hierarchical variable. For example, a geographical description can be made with the town or with the county or the country. The taxonomical variable describing the location will use any level of the description (town, county or country). In the data set we can find instances with a location given by a town name and other ones with a county or country name. From the hierarchical description, we get a set of ordered values by using a tree traversal (either depth-first or width-first). Let us show the results on a very simple example of geographical location. The location is defined by the binary tree described in figure 6. The leaves correspond to town, and the upper levels to county and country. In the data set, the location attribute can take any value of this tree (except the root value). An example of such a data set is given in table 1, the a priori class has two possible values: 1 and 2. The two columns on the left correspond to the original data, the two columns in the middle are the same data set sorted according to a depth-first traversal of the tree, and the two columns on the right are the same data set sorted according to a breadth-first traversal of the tree.

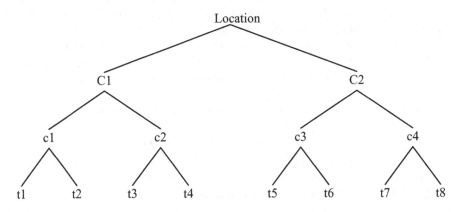

Fig. 6. Hierarchical description of the geographical location

Once the data have been sorted (whatever the tree traversal is), a taxonomical variable can be seen as an interval variable. When the variable is not a leaf of the tree (for example C1 or c3 in figure 6), it is graphically equivalent to the interval made of all the leaves of the corresponding sub-tree (C1=[t1,t4] and c3=[t5,t6]). We will use exactly the same representation as for interval-valued data.

Table 1. An example of taxonomical dataset, unordered on the left, ordered according to a depth first tree traversal in the center and a breadth first tree traversal on the right.

Loc.	Class		Loc. (depth)	Class		Loc. (breadth)	Class
t1	1		t1	1		C1	2
t2	2		c1	2		c1	2
t3	1		t2	2		c3	2
t3	1		C1	2		t1	1
c1	2		t3	1		t2	2
C1	2		t3	1		t3	1
t5	1		t5	1		t3	1
c3	2		c3	2		t5	1
t7	1		t7	1		t7	1

Here again, the way PBC and CIAD are used is exactly the same as for interval or continuous data. There is an order for the taxonomical data, it is used in the first step of the PBC algorithm, to sort the data according to the attribute value.

4 Some Results

First of all, we must underline that as far as we know, there is no other decision tree algorithm able to deal with continuous, interval and taxonomical data and there are no available interval-valued data sets in existing machine learning repositories. We present in this section some of the results we have obtained and we start with the description of the data sets we have created. We have used existing data sets with continuous variables to create the interval-valued data sets. The first data set used is the well-known iris data set from the UCI Machine Learning Repository (150 instances, 4 attributes (the petal and sepal length and width) and 3 classes (the iris type)). First, a new attribute has been added to data set: the petal surface. Then the data set has been sorted according to this new attribute (it nearly perfectly sorts the iris types) and we have computed (for each attribute) the minimum and the maximum values of each group of five consecutive items. And so we obtain a data set made of four interval-valued attributes and 30 items (10 for each class).

The second data set we have created is an interval-valued version of the Shuttle data set (43700 instances, 9 continuous attributes and 7 classes). We have used a clustering algorithm to create groups of instances for each class. Then we use the minimal and maximal values of each cluster according to each dimension to create interval-valued attributes.

Once the data are displayed, we perform the interactive creation of the decision tree. The accuracy obtained on the training set is 100% with a ten leaves tree (99.7% on the test set). On the continuous data set, the accuracy obtained with CIAD was 99.9% with nine leaves.

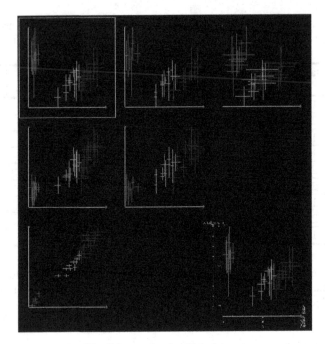

Fig. 7. Interval-valued Iris data set

These first results show the accuracy is nearly the same with the original data set and the interval-valued one we have created. This new kind of algorithm allow us to deal with potentially very large data sets because we do not use the data themselves but a higher-level representation of the data. A data set such as Forest Cover Type (with only 5810 interval-valued items) can easily be treated with a graphical interactive decision tree construction algorithm like CIAD or PBC.

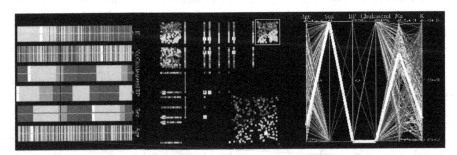

Fig. 8. Three linked representations of the same dataset: on the left, PBC, in the center CIAD and the parallel coordinates on the right

Before concluding, some words about the implementation. All these tools have been developed using C/C++ and three open source libraries: OpenGL, Open-Motif and Open-Inventor. OpenGL is used to easily manipulate 3D objects, Open-Motif for the graphical user interface (menus, dialogs, buttons, etc.) and Open-Inventor to manage the 3D scene. These tools are included in a 3D environment, described in [Poulet, 2002], where each tool can be linked to other tools and be added or removed as needed. Figure 8 shows an example with CIAD set of 2D scatter plot matrices, PBC bar charts and parallel coordinates. The element selected in a bar chart appeared selected too in the set of scatter plot matrices and in the parallel coordinates [Inselberg, 1985]. The software program can be run on any platform using X-Window, it only needs to be compiled with a standard C++ compiler. Currently, the software program is developed on SGI O2 and PCs with Linux.

5 Conclusion and Future Work

In this paper we have presented new interactive classification tools able to deal with interval-valued and taxonomical data. With continuous data, the accuracy obtained by interactive algorithms is not significantly different from the accuracy obtained with automatic algorithms. The main advantages of interactive classification tools are:

- they improve the result comprehensibility and the confidence in the results (because the user has taken part in the model construction),
- they can exploit human capabilities in graphical analysis and pattern recognition,
- we can take into account the domain knowledge, if the user is a data specialist.

The main drawback of this kind of approach and of several visualization methods is the difficulty to deal with large dataset. This drawback is overcome by using a higher-level data representation. These new versions are able to deal with interval-valued and taxonomical data and so with potentially very large datasets with the same accuracy as with usual quantitative data. Another possibility, we have already started to work on [Poulet, 2004] is to use simultaneously automatic and interactive methods in a cooperative way.

A forthcoming improvement will be to deal with histogram variables, we can use the order defined on these variables to sort the histogram variables in the first step of the PBC algorithm. Once the attributes values are sorted, we can use exactly the same mechanism to construct a decision tree on any mix of continuous, interval, taxonomical and histogram variables.

The use of symbolic data can also be extended to other supervised classification algorithms like Support Vector Machine or other related kernel methods [Do and Poulet, 2005].

Finally, we have only given a potential solution to reduce the number of items of the datasets (rows of the databases) but some applications, like text mining or bioinformatic, deal with very large number of dimensions (columns of the databases). We must find another approach to deal with this kind of datasets with graphical interactive classification methods.

References

[Aggarwal, 2001] Aggarwal C., *Towards Effective and Interpretable Data Mining by Visual Interaction*, in SIKDD Explorations 3(2), 11-22, accessed from www.acm.org/ sigkdd/ explorations/.

[Ankerst, 2000] Ankerst M., *Visual Data Mining*, PhD Thesis, Faculty of Mathematics and Computer Science, Univ. of Munich, 2000.

[Ankerst et al, 2001] Ankerst M., Ester M., Kriegel H-P., *Toward an Effective Cooperation of the Computer and the User for Classification*, in proc. of KDD'2001, 179-188.

[Asseraf et al, 2004] Asseraf M., Mballo C., Diday E., *Binary decision trees for interval and taxonomical variables*, in A Statistical Journal for Graduate Students, Presses Académiques de Neuchâtel, 5(1):13-28, 2004.

[Blake and Merz, 1998] Blake C., Merz C., UCI Repository of machine learning databases, [http://www.ics.uci.edu/~mlearn/MLRepository.html]: University of California Irvine, Department of Information and Computer Science.

[Bock and Diday, 2000] Bock H.H., Diday E, *Analysis of Symbolic Data: : Exploratory Methods for Extracting Statistical Information from Complex Data*, Springer-Verlag, 2000.

[Carr et al, 1987] Carr D., Littlefield R., Nicholson W., Littlefield J., Scatterplot Matrix Techniques for Large N, Journal of the American Statistical Association 82(398), 424-436, 1987.

[Ciampi et al, 2000] Ciampi A., Diday E., Lebbe J., Périnel E., Vignes R., *Growing a tree classifier with imprecise data*, in Pattern Recognition Letters 21:787-803, 2000.

[Do and Poulet, 2005] Do T-N., Poulet F., *Interval Data Mining with Kernel Methods and Visualization*, in proc. of ASMDA'2005, XIth International Symposium on Applied Stochastic Models and Data Analysis, Brest, France, May 2005, 345-354.

[Fayyad et al, 1996] Fayyad U., Piatetsky-Shapiro G., Smyth P., Uthurusamy R., *Advances in Knowledge Discovery and Data Mining*, AAAI Press, 1996.

[Han and Cercone, 2001] Han J., Cercone N., *Interactive Construction of Decision Trees*, in proc. of PAKDD'2001, LNAI 2035, 575-580, 2001.

[Inselberg, 1985] Inselberg, A., *The plane with parallel coordinates*, in Special Issue on Computational Geometry, 1:69-97, 1985.

[Keim et al, 1995] Keim D., Kriegel H-P., Ankerst, M., *Recursive Pattern: A Technique for Visualizing Very Large Amount of Data*, in proc. of Visualization'95, Atlanta, USA, 1995, 279-286.

[Mballo and Diday, 2004] Mballo C., Diday E., *The criterion of Kolmogorov-Smirnov for binary decision tree: application to interval valued variables*, in proc. of ECML/PKDD'2004 Workshop on Symbolic and Spatial Data Analysis, P.Brito and M.Noirhomme-Fraiture Eds, 79-90, 2004.

[Poulet, 2001] Poulet F., *CIAD: Interactive Decision Tree Construction*, in proc. of XXXIIIe Journées de Statistiques, Nantes, May 2001 (in french).

[Poulet, 2002] Poulet F., *Full-View: A Visual Data-Mining Environment*, in IJIG: International Journal of Image and Graphics, Vol.2, N.1, Jan. 2002, pp.127-144.

[Poulet, 2004] Poulet F., *SVM and Graphical Algorithms: A Cooperative Approach*, in proc. of IEEE ICDM'04, International Conference on Data Mining, Brighton, UK, Nov.04, 499-502.

[Shneiderman, 2002] Schneiderman B., *Inventing Discovery Tools: Combining Information Visualization with Data Mining*, in Information Visualization 1(1), 5-12, 2002.

[Ware et al, 2001] Ware M., Franck E., Holmes G., Hall M., Witten I., *Interactive Machine Learning: Letting Users Build Classifiers*, in International Journal of Human-Computer Studies (55), 281-292, 2001.

[Wong, 1999] Wong P., *Visual Data Mining*, in IEEE Computer Graphics and Applications, 19(5), 20-21, 1999.

Using Biclustering for Automatic Attribute Selection to Enhance Global Visualization

Ahsan Abdullah[1,2] and Amir Hussain[2]

[1] National University of Computers & Emerging Sciences, Pakistan
ahsan@nu.edu.pk
[2] University of Stirling, Stirling, Scotland, UK
ahu@cs.stir.ac.uk

Abstract. Data mining involves useful knowledge discovery using a data matrix consisting of records and attributes or variables. Not all the attributes may be useful in knowledge discovery, as some of them may be redundant, irrelevant, noisy or even opposing. Furthermore, using all the attributes increases the complexity of solving the problem. The Minimum Attribute Subset Selection Problem (MASSP) has been studied for well over three decades and researchers have come up with several solutions In this paper a new technique is proposed for the MASSP based on the crossing minimization paradigm from the domain of graph drawing using biclustering. Biclustering is used to quickly identify those attributes that are significant in the data matrix. The attributes identified are then used to perform one-way clustering and generate pixelized visualization of the clustered results. Using the proposed technique on two real datasets has shown promising results.

1   Introduction

Every day business managers, biologists, analysts, agriculture scientists etc. are confronted with large datasets to be explored and analyzed-effectively and quickly. The number and complexity of such datasets is growing by the day. Extracting knowledge from these datasets have two conflicting objectives (i) present the data using all the relevant attributes in an intuitive and easy to understand way for non technical users, such that (ii) the data retains all the important and relevant relationships. Thus there is a need for clustered visual representation of useful and relevant data for visual data mining. For this purpose the screen or computer display is the working area, where the smallest unit of information is a pixel. Thus pixelization is a natural choice for visual data mining.

Thus an important and critical issue in the data preprocessing step of the overall Knowledge Discovery in Databases (KDD) is the selection of useful and relevant attributes allowing data mining algorithm to discover useful patterns effectively and quickly. It is typical of the data mining domain that the number of attributes is very large and not all of them may be effective in the overall knowledge discovery process. A possible way is to use the domain expert's knowledge for the identification of most suitable attributes, and then run the data mining algorithms and hope that they give

P.P. Lévy et al. (Eds.): VIEW 2006, LNCS 4370, pp. 35–47, 2007.

the most useful results with the selected attribute set. However, deciding 'suitability' is an issue, and only quantifying and relating it to some measure can make it automatic and objective.

Among a number of factors affecting the success of a data mining algorithm is data quality. No matter how "intelligent" a data mining algorithm is, it will fail in discovering useful knowledge if applied to low-quality data [5]. The irrelevant, redundant, noisy and unreliable data makes the knowledge discovery task difficult. Minimum Attribute Subset Selection Problem (MASSP) is the process of identifying and removing as much of the irrelevant and redundant information as possible [11]. A carefully chosen subset of attributes improves the performance and efficacy of a variety of algorithms [3]. A relevant feature is neither irrelevant nor redundant to the target concept; an irrelevant feature does not affect the target concept in any way, and a redundant feature does not add anything new to the target concept [12].

Subset selection problem is also known as dimensionality reduction, as it is the presentation of high dimensional patterns in a low dimensional subspace based on a transformation which optimizes a specified criterion in the subspace [13]. Dimensionality reduction is most useful when each of the n inputs carry information relevant to classification i.e. need a *global view* of data. On the other hand, subset selection is appropriate when it is possible to completely discard some attributes (i.e. irrelevant attributes) with low probability of misclassification [13] i.e. need a *local view* of data.

The MASSP involves finding a 'good' set of attributes under some objective function that assigns numeric measure of quality to the patterns discovered by the data mining algorithm. It has long been proved that the classification accuracy of Machine learning algorithms is not monotonic ((i.e., a subset of features should not be better than any larger set that contains the subset [12]) with respect to the addition of features. Some possible objective functions for classification could be the prediction accuracy, misclassification cost, pattern complexity, minimal use of input attributes [6] etc. In this paper, the objective function is the crossing minimization of the bipartite graph corresponding to the discretized data matrix. The clusters found are subsequently quantified using a Figure-of-Merit (FoM) which is the average of the standard deviation (Std-Dev) of the column values.

Attribute subset selection problem has been defined in a number of ways by different researchers. However, though differing in expression, the definitions resemble in intuition. In [12] four definitions have been listed that are conceptually different and cover a wide range and are as follows:

- *Idealized:* Find the minimally sized feature subset that is necessary and sufficient to the target concept.
- *Classical:* Select a subset of P features from a set of Q features, P < Q, such that the value of a criterion function is optimized over all subsets of size P.
- *Improving Prediction accuracy:* The aim of feature selection is to choose a subset of features for improving prediction accuracy or decreasing the size of the structure without significantly decreasing prediction accuracy of the classifier built using only the selected features.

- *Approximating original class distribution:* The goal of feature selection is to select a small subset such that the resulting class distribution, given only the values for the selected features, is as close as possible to the original class distribution given all feature values.

1.1 Advantages of Solving MASSP

Some of the advantages of using the subset of attributes are as under [6, 8, 11]:

- A reduction in the cost of acquisition of the data.
- It can speed up the data mining task by reducing the amount of data that needs to be processed.
- Reducing the dimensionality reduces the size of hypothesis space and allows algorithms to operate faster and more effectively.
- Reducing the number of available attributes usually reduces the number of attributes in the discovered patterns, enhancing ease of understanding for the decision maker.
- Most data mining algorithms can get confused by the presence of irrelevant and even relevant but harmful attributes, thus removing those attributes is expected to lead to better quality results.

2 Background

Data is typically stored in a table or a database, also called as a data matrix. From data matrix similarity (or dissimilarity) matrices are created. Clustering of the similarity (or dissimilarity) matrix is called one-way clustering, which gives a *global view* of the data. Simultaneous clustering of the rows and columns of the data matrix is called two-way clustering, which gives a *local view* of the data. For one-way clustering choice of columns (or attributes) is critical, as this may actually hide, or do not completely display the clusters present. There are techniques based on supervised [7] and unsupervised classification [19] for attribute selection, in this paper; a new technique is proposed for the MASSP using biclustering, and biclustering is achieved by crossing minimization [1].

Let the data matrix be denoted by S and its discretized version be denoted by S_B. Let the bipartite graph corresponding to S_B be denoted by G_B, such that one bipartition of G_B i.e. BOT (bottom) consists of vertices that correspond to the rows (or records) and the second bipartition i.e. TOP (top) to consist of vertices that correspond to the columns (or attributes). Fig-1(a) shows a discretized data matrix i.e. S_B consisting of 16 rows and 7 attributes. There will be edges between those vertices of G_B, for which there is a non-zero value in S_B. Fig-1(b) shows the G_B corresponding to S_B. Discretization is critical to the proposed solution, as it reduces the complexity of the problem i.e. converts a weighted clique into a binary i.e. 1-0 graph of lower density. Discretization is achieved by partitioning continuous variables or attributes into discrete values or categories, typically using column median or average. Working with S will force considering each and every edge of the un-discretized graph, hence a highly undesirable $\Omega(n^2)$ time complexity. Working with S_B reduces the time complexity to $O(K)$, where K is the number of 1's in S_B. Thus, the viable way out is the discretization of S, leading to S_B. Note that even algorithms with quadratic time complexities are unacceptable for most KDDM (Knowledge Discovery by Data Mining) applications according to Fayyad and Uthurusamy [18].

3   Related Works

Researchers have used many techniques and methods for solving the MASSP. As a consequence it is difficult to compare all these techniques and indeed such a comparison is outside the scope of this paper. However, an effort is made to give a broad overview of the work done. Attribute selection algorithms can generally be classified into three categories based on whether or not attribute selection is done independently of the learning algorithm used to construct the classifier: filter, wrapper and embedded approaches. They can also be classified into three categories according to the search strategy used: exhaustive search, heuristic search and randomized search [10].

3.1   Filter Methods

These methods act as a filter in the data mining process, to remove possible irrelevant attributes before the application of the data mining algorithms. The methods work independently of the data mining algorithm (induction algorithm) [6]. In such methods, the goodness of an attribute subset can be assessed with respect to the intrinsic properties of data. The statistics literature proposes many measures for evaluating the goodness of a candidate attribute subset. Examples of statistical measures used are information gain and correlation [4] etc.

3.2   Wrapper Methods [6]

In this method the attribute subset selection algorithm exists as a wrapper around the data mining algorithm and the method that evaluates the results. The attribute subset selection algorithm conducts a search for a good subset using the mining algorithm, which itself is a part of the function that evaluates an attribute subset. In case of classification problems, the evaluation of the discovered classification model is used as an estimate of the accuracy through cross validation. The essential characteristic of the wrapper method is the usage of same data mining algorithm and the evaluation method that will be subsequently used in the data mining process. No knowledge of the induction algorithms is necessary, except testing the resulting patterns on the validation sets. Many algorithms have been suggested in the statistics literature for finding a good subset of attributes under various assumptions. However, these assumptions do not hold for most of the data mining algorithms, hence heuristic search is used mostly.

3.4 MASSP as a Search Problem

MASSP can be considered as an optimization problem which involves searching the space of possible attribute subsets and identifying the ones that are optimal or nearly optimal with respect to a performance measure [10]. The search space consists of $O(2^m)$ possible solutions, here m is the number of attributes in the data matrix. Many search techniques have been used for solving the MASSP intelligently, but exhaustively searching the entire space is prohibitive especially when the number of attributes is large [13]. For example, randomized, evolutionary, population based search techniques and Genetic Algorithms (GAs) have long been used in the MASSP solving process. GAs needs crossover and mutation operators to make the evolution possible, whose optimal selection is a hard problem.

3.4 Embedded [8]

In this approach the learning algorithm prefers some attributes instead of the others, and in the final classification model possibly not all of the available attributes are present. For example, some induction algorithms like partitioning and separate–and-conquer methods implicitly select attributes for inclusion in a branch or rule by giving preference to those attributes that appear less relevant, consequently some of the attributes never get selected. On the other hand, some induction algorithms like Naïve-Bayes include all the presented attributes in the model when no attribute selection method is executed. Conceptually the filter and the wrapper methods are located one abstraction level above the embedded approach, because they perform an attribute selection for the final classifier, apart from the embedded selection done by the learning algorithm itself.

3.5 Comparative Study of Attribute Subset Selection Methods

Some good work on reviewing the feature subset selection methods can be found in [12], [11]. The survey by [12] covers feature selection methods starting from the early 1970's to 1997. A framework is suggested that helps in finding the unexplored combinations of generation procedures and evaluation functions. Thus 16 representative feature selection methods based on the framework are explored for their strengths and weaknesses regarding issues like attribute types, data noise, computational time complexity etc.

The work by [11] presents a benchmark comparison of six major attribute selection methods: Information Gain (IG), Relief (RLF), Principal Components (PC), Correlation –based Feature Selection (CFS), Consistency (CNS) and Wrapper (WRP), on 14 well known and large benchmark datasets (two containing several hundreds of features and the third over a thousand features) for supervised classification. The benchmark shows that in general, attribute selection is beneficial for improving the performance of common learning algorithms and suggests some general recommendations.

4 Crossing Minimization and Attribute Identification

The aim of this section is to give (i) a brief background of the crossing minimization paradigm, (ii) working of the technique that performs biclustering using crossing minimization, (iii) a simple example demonstrating that technique and (iv) finally explain how attribute selection using biclustering saves time.

4.1 Crossing Minimization

The crossing minimization problem has been studied for over two decades, and its two variants i.e. one-layer and two layers are known to be NP-Complete problems - see Garey and Johnson [14]. There are basically three types of crossing minimization heuristics. The first type are the classical ones that do not count the crossings, hence are very fast and yet give good results such as BaryCenter Heuristic BC by Sugiyama et al [9] and Median Heuristic (MH) by Eades and Wormald [15]. Another non-crossing counting heuristic is MinSort (MS) by Abdullah [2] used in this paper. Then there are the crossing counting heuristics that use a diverse set of approaches, such as greedy, tabu-search, simulated annealing, genetic algorithms, etc. and work by counting the number of crossings and, hence, carry the counting overhead. Lastly, there are meta-heuristics i.e. use of the classical heuristics to generate the initial arrangement of vertices, and then improve upon it using other heuristics. For a detailed study see Marti and Laguna [16].

4.2 Working of the Proposed Attribute Identification Technique

For ease of understanding, the proposed biclustering technique [1] used for identification of attributes is presented below in a rather oversimplified form, as a four-step procedure:

1. Discretize S to obtain S_B.
2. Run a CMH on G_B corresponding to S_B to get a crossing reduced result G^+_B.
3. Extract biclusters from G^+_B.
4. Analyze biclusters.

The bicluster extraction method is described in more detail as follows:

1. Start with a null bicluster scanning from the top-left corner of S_B.
2. Incrementally grow the bicluster (say) along x-axis using a window of size Ψ.
3. Grow the bicluster as long as the density contributed by the incremental growth is greater than the density of S_B.
4. If the density of Ψ becomes equal to the density of the bin-matrix, then start growing the bicluster in the other direction (y-axis).
5. If the contributed density is less in both directions, then freeze the bicluster or else continue until the entire S_B is exhausted.
6. Repeat the above steps (1-5), starting from the bottom-right corner of S_B.
7. Combine the results, display and list the biclusters.

Example-1

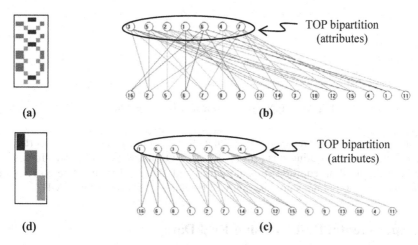

Fig. 1. (a) S_B. (b) G_B consisting 288 crossings. (c) G^+_B consisting of 66 crossings. (d) S^+_B.

Figure-1(a) shows a 16×7 discretized S_B and Fig-1(b) shows the corresponding G_B. Attributes of the data matrix correspond to the TOP bipartition of the bipartite graph. After crossing minimization Fig-1(c) is obtained and the corresponding optimally arranged S^+_B is shown in Fig-1(d). The subset of attributes of interest have been identified as (1, 6), (3, 5, 7) and (2, 4) in Fig-1(c). For ease of comprehension an example of non-overlapping attributes is presented, but the proposed technique works for overlapping attributes also.

One-way clustering is typically performed by creating a pair-wise similarity or dissimilarity matrix using measures such as Pearson's correlation. Assuming that the data matrix consists of n rows and m columns, creating the similarity matrix takes $O(n^2m)$ time. This is subsequently followed by the clustering algorithm, however, the bottleneck is the time spent in creating the similarity matrix. Our proposed technique reduces the number of attributes m by spending $O(nm)$ time using biclustering. Once the biclusters are extracted, the same are analyzed to identify the attributes which will be used to perform clustering as done previously in Example-1.

Example-2
In this example working of the MS crossing minimization heuristic is demonstrated for a simple bi-graph.

Figure-2(a) shows a bi-graph with 5 crossings. First the vertices in the fixed layer (indicated by the arrow) are assigned ordering i.e. 1, 2, and 3. Subsequently the orderings for the chengeable layer are generated based on the minimum value resulting in reordering of vertices to 2, 3 and 1 as shown in Figure-2(b) BOT bipartition. This process is repeated till equilibrium is reached i.e. no further chenge in orderings. The final non-optimal output is shown in Figure-2(c); with two crossings.

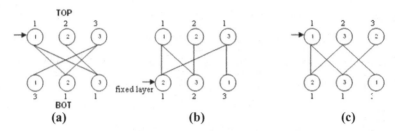

Fig. 2. (a) Input. (b) Intermediate result. (c) Output.

For a G_B numbering the vertices in the TOP and subsequently sorting vertices in the BOT, and then repeating this process for the BOT has the effect of two forces that are alternated until an equilibrium state is reached. An equilibrium state corresponds to the final output that does not change with further application of that CMH.

5   Experimental Results Using Real Data

The aim of this section is to demonstrate the working of the proposed technique using two datasets i.e. the glass data set and the wine data set, both publicly available at http://www.ics.uci.edu/~mlearn/. In the first experiment the glass data set is used to demonstrate the concept of the proposed technique. The data set consist of 214 records, nine attributes (excluding the class attribute) and seven classes. In the said data set the records are already ordered by their class. A simple analysis would involve creating a pair wise similarity matrix of the given data set, and color coding it by assigning bright green color to the value of 1 and red color to the value of -1. Intermediate values are assigned shades of the corresponding color, with black color being assigned to a correlation of 0. The resulting color coded similarity matrix for the glass data set is shown in Fig-3(a), as can be seen that nothing is visible.

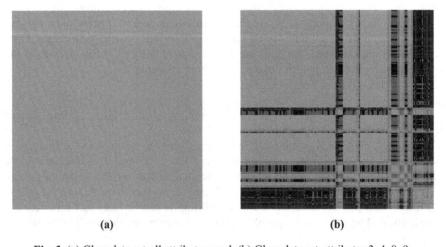

Fig. 3. (a) Glass data set-all attributes used. (b) Glass data set-attributes 3, 4, 8, 9.

For selecting the attributes so as to enhance visualization, biclustering was performed on the glass data set. Based on the results of biclustering, the minimum attributes of significance were identified to be 3, 4, 8 and 9 i.e. Na, Mg, Ca and Ba. Using these attributes the similarity matrix was again generated, and the color coded matrix is shown in Figu-3(b). Observe the significantly enhanced visualization with lots of interesting relationships and clusters distinctly visible.

Now the wine data set is used for a detailed experiment which will also demonstrate how biclustering is used for attribute selection and minimization. The data set consists of 178 rows, 13 attributes and three clusters. For the ease of understanding, a relatively small data set has been selected that can easily be displayed on a single screen of 600×800 pixel resolution. The 13 attributes of the dataset are (1) 'Alcohol', (2) 'Malic', (3) 'Ash', (4) 'Alcalinity', (5) 'Magnesium', (6) 'Phenols', (7) 'Flavanoids', (8) 'Nonflavanoids', (9) 'Proanthocyanins', (10) 'Color', (11) 'Hue', (12) 'Dilution', and (13) 'Proline'.

Fig-4(a) shows the corresponding color coded clustered similarity matrix, some clustering is visible, but visualization deteriorates when real valued similarity matrix is discretized, as shown in Fig-4(b).

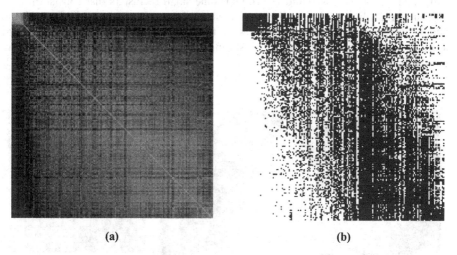

(a)   (b)

Fig. 4. (a) Clustered similarity matrix of size 178 × 178 using all 13 attributes. (b) Binary version of Fig-4(a).

To get the subset of attributes for enhancing clustering, biclustering of the wine dataset is performed using the indigenous crossing minimization technique [1]. Firstly the given data matrix is normalized and color coded it, such that bright green color is assigned to value of 1 and black to value of 0, intermediate values are assigned shades of green. The color coded unclustered data matrix rotated by 90° is shown in Fig-5(a). Secondly biclustering is performed using the discretized binary data matrix, the color coded biclustered data matrix is shown in Fig-5(b), while the discretized version of Fig-5(b) with biclusters extracted (red rectangles) is shown in Fig-5(c).

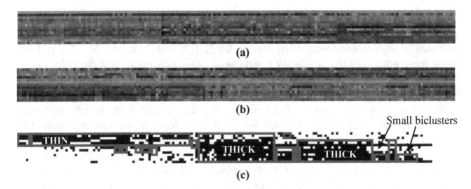

Fig. 5. (a) Color coded input data matrix of size 178 ×13 rotated by 90°. (b) Color coded output biclustered data matrix. (c) Binary biclustered data matrix with biclusters shown by red rectangles.

From Fig-5(c) three significant biclusters are identified titled "thin" of size 41×3 and "thick" of sizes 24×10 and 19×10 and some small biclusters that consist of three columns or attributes. The small biclusters share three columns or attributes with the "thick" biclusters, and no columns with the "thin" bicluster. Thus two sets of attributes are obtained through biclustering and subsequently used for one-way clustering i.e. set-1 consisting of attributes 2, 4, 8 and set-2 consisting of attributes 3, 5, 10.

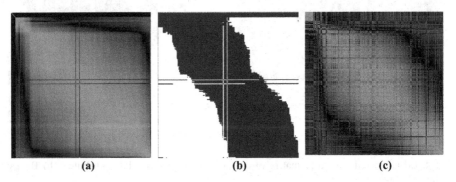

Fig. 6. (a) Input similarity matrix of size 178 × 178 using set-1. (b) Binary version of Fig-6(a). (c) Input similarity matrix of size 178×178 using attributes 3, 8, and 11.

Using set-1 one-way clustering of the wine dataset is performed, and the results are shown in Fig-6(a). Observe the enhanced visualization with negative correlations also visible and very well defined pixelized binary visualization as shown in Fig-6(b).

5.1 Quantitative Analysis of Results

The aim of this section is to quantitatively analysis the one-way clustering results with all the attributes used and with fewer attributes used. For the purpose of quantification, a simple yet effective Figure-of-Merit or FoM is defined. The FoM being the average of the standard deviation (Std_Dev) of the columns values of records or rows included in each cluster. Clusters with rows of highly similar values will have a low FoM and is desirable.

For the wine dataset, the FoM of the unclustered input is 26.17. Quantitative analysis of the results is performed by noting the density and the FoM of each cluster. In the context of the binary data matrix, clusters with high density apparently seem to correspond to good clustering, but this may not always be true. The reason being, it hides the actual column values which may all be above the median value of the entire column of the input data matrix, but all different within the part of the column included in the corresponding cluster.

Table 1 shows the results of using all 13 attributes of the wine data set with high values of FoM. Table 2 shows the results of using set-1 and set-2 with attributes identified using biclustering through crossing minimization. Observe the significant reduction (or improvement) in the FoM.

Table 1. Results of using all 13 attributes of the wine dataset for clustering

All 13 attributes	Cluster size	Density	FoM
	16 × 16	1.00	17.93
	40 × 40	0.90	17.6
	36 × 36	0.83	15.4

Table 2. Results of using set-1 and set-2

Three attributes of set-1 i.e. 2, 4 and 8	Cluster size	Density	FoM
	82 × 82	0.79	1.59
	89 × 89	0.87	1.42
Three attributes of set-2 i.e. 3, 5 and 10	81 × 81	0.76	4.72
	27 × 27	0.85	4.28
	44 × 44	0.93	3.82
	16 × 16	0.90	3.92

A trivial way of improving FoM is to use attributes with low Std_Dev. This could be done by calculating the Std_Dev of each attribute in O(n) time and then sorting them based on their Std_Dev in O(m log m) time. Subsequently the attributes with low Std_Dev are selected to perform one-way clustering. This technique does not take into account any relationships between the attributes (as our proposed technique does) and attributes are treated as independent of each other. Based on this trivial approach the top three attributes in the wine dataset with ascending Std_Dev are 8, 11 and 3. Note the contradiction as attribute 3 belongs to set-1 while attribute 8 belongs to the set-2. Using these attributes to perform clustering, resulted in numerous negative correlations or red colored pixels visible in the clustered similarity matrix and shown in Fig-6(c). The paradox is that while trying to naively improve the FoM, the trivial approach ends up using those attributes which don't collectively support clustering.

6 Conclusions

Minimum Attribute subset selection problem (MASSP) has generated lot of interest, because of the increase in the number of attributes in datasets i.e. into hundreds. Lot of work has been done to address the MASSP problem in the context of data mining, but biclustering based on the crossing minimization paradigm is a first effort of its kind. It is observed that using the proposed technique not only visually better results are obtained, but the quality of those results was also better quantitatively. It is also observed that using a naïve approach of picking attributes with high similarity may actually deteriorate the cluster quality. The proposed technique takes into account the relationships between records and attributes and thus gives good results. It would be interesting to use other biclustering techniques [17] for solving the ASSP problem and compare their results.

References

1. A. Abdullah and A. Hussain, "A new biclustering technique based on crossing minimization", to appear in the *Neurocomputing* Journal, 2006.
2. A. Abdullah and S. Brobst, "Clustering by recursive noise removal", Proc. Atlantic Symposium on Computational Biology and Genome Informatics", North Carolina, pp. 973-977, 2003.
3. C. Traina, L. Wu, A. Traina, and C. Faloutsos, "Fast Feature Selection Using Fractal Dimension " XV Brazilian Symposium on Databases (SBBD), Paraiba, Brazil, October, 2000
4. E. F. Badjio and F. Poulet, "Dimension Reduction for Visual Data Mining", ESIEA Recherche, Parc Universitaire de Laval-Change, 38 Rue des Docteurs Calmette et Guerin, 53000 Laval, France, 2005.
5. G. L. Pappa, A. A. Freitas and C. A. A. Kaestner, "A multiobjective Genetic Algorithm for Attribute Selection", *Proc. 16th Brazilian Symp. on Artificial Intelligence (SBIA-2002)*, Lecture Notes in Artificial Intelligence 2507, pages 280-290. Springer-Verlag, November 2002.
6. G. H. John "Enhancements to the Data Mining Process", PhD Dissertation, Stanford University, 1997.
7. H. Liu and H. Motoda, "Feature selection for knowledge discovery and data mining", in Kluwer International Series in Engineering and Computer Science, Secs, 1998.
8. I. Inza, P. Larranga, R. Etxeberria and B. Sierra, "Feature Subset Selection by Bayesian networks based optimization", Artificial Intelligence, vol. 123, no. 1-2, pp. 157-184, 2000.
9. K. Sugiyama, S. Tagawa and M. Toda, Methods for Visual Understanding of Hierarchical Systems. IEEE Trans. Syst. Man Cybern., SMC-11(2):109-125, 1981.
10. L. Boudjeloud and F. Poulet, "Attribute Selection for High Dimensional Data Clustering", ESIEA Recherche, Parc Universitaire de Laval-Change, 38 Rue des Docteurs Calmette et Guerin, 53000 Laval, France, 2005.
11. M. A. Hall and G. Holmes, "Benchmarking Attribute Selection Techniques for Discrete Class Data Mining", IEEE Trans. on knowledge discovery and Data Engineering, VOL. 15, NO. 3, MAY/JUNE 2003.
12. M. Dash , H. Liu, "Feature Selection for Classification", Intelligent data Analysis, vol.1, no.3, 1997.

13. M. Dong, "A New Measure of Classifiability and Its Applications", PhD Dissertation Department of Electrical & Computer Engineering and Computer Science of the College of Engineering, University of Cincinnati, 2001.
14. M. R. Garey and D. S Johnson, Crossing number is NP-Complete, SIAM J. Algebraic Discrete Methods, 4(1983), 312-316, 1983.
15. P. Eades, N. Wormald, The Median Heuristic for Drawing 2-layers Networks, Tech. Report 69, Department of Computer Science, University of Queensland, Brisbane, Australia, 1986.
16. R. Marti and M. Laguna, Heuristics and Meta Heuristics for 2-layer Straight Line Crossing Minimization, Discrete Applied Mathematics, Vol. 127 - Issue 3, pp. 665 – 678, 2001.
17. S. C. Madeira and A. L. Oliveira, "Biclustering Algorithms for Biological Data Analysis: A Survey", IEEE/ACM Transactions on Computational Biology and Bioinformatics, 2004.
18. U. Fayyad, and R. Uthurusamy, Data Mining and Knowledge Discovery in Databases, Comm. ACM, vol. 39, no. 11, pp. 24-27, 1996.
19. Y. Kim, W. N. Street, and F. Menczer, "Evolutionary model selection in unsupervised learning", in volume 6, pages 531–556. IOS Press, 2002.

Pixelisation-Based Statistical Visualisation for Categorical Datasets with Spreadsheet Software

Gaj Vidmar

University of Ljubljana, Faculty of Medicine, Institute of Biomedical Informatics
Vrazov trg 2, SI-1000 Ljubljana, Slovenia
gaj.vidmar@mf.uni-lj.si
http://www.mf.uni-lj.si/ibmi-english

Abstract. A heat-map type of chart for depicting large number of cases and up to twenty-five categorical variables with spreadsheet software is presented. It is implemented in Microsoft® Excel using standard formulas, sorting and simple VBA code. The motivating example depicts accuracy of automated assignment of MeSH® descriptor headings to abstracts of medical articles. Within each abstract, predicted support for each heading is ranked, then for each heading actually assigned/non-assigned by human specialist (depicted by black/white cell), high/low support is depicted on nine-point two-colour scale. Thus, each case (abstract) is depicted by one row of a table and each variable (heading) with two adjacent columns. Rank-based classification accuracy measure is calculated for each case, and rows are sorted in increasing accuracy order downwards. Based on analogous measure, variables are sorted in increasing prediction accuracy order rightwards. Another biomedical dataset is presented with a similar chart. Different methods for predicting binary outcomes can be visualised, and the procedure is easily extended to polytomous variables.

1 Introduction

The rapid progress in data visualisation in the recent years has not left categorical data behind. This is exemplified in the literature [1] and in software, such as the tools from the RoSuDa group from the University of Augsburg (MANET, Mondrian and other programs[1]), Treemap[2] and its commercial extensions (e.g., Panopticon's products[3]), and various other products (like Miner3D[TM4]). Through software, often publicly available as code for freely available or commercial statistical packages (mainly R, S-Plus® or SAS®), or add-ins (like VisuLab for Microsoft® Excel[5]), older methods, such as permutation matrices [2] and mosaic plots [3, 4], have found more use, and newer methods have found wide applications in specialised areas, e.g., heat maps [5, 6] in gene microarray data analysis.

---

[1] Publicly available from http://stats.math.uni-augsburg.de
[2] http://www.cs.umd.edu/hcil/treemap
[3] http://www.panopticon.com
[4] http://www.miner3d.com
[5] http://www.inf.ethz.ch/personal/hinterbe/Visulab

P.P. Lévy et al. (Eds.): VIEW 2006, LNCS 4370, pp. 48–54, 2007.
© Springer-Verlag Berlin Heidelberg 2007

In parallel, spreadsheets have gradually found their way into the world of mathematics and statistics. Despite limitations and shortcomings, mainly in terms of numerical algorithms and missing data handling [7], which can take considerable effort and expertise to overcome, they have proven to be a valid tool in mathematics and statistics education [8], as well as in statistical practice[6].

One of the connecting points between the two emphasised areas of development is that various visualisations of categorical data can be implemented with tables, and the natural means for such purpose are electronic spreadsheets. The most widespread of them is Microsoft[®] Excel, which also has an excellent public knowledge base consisting of the official newsgroup[7] and the websites of the leading MVP[8] awardees, such as F. Cinquegrani[9], J. Peltier[10], D. McRitchie[11] and J. Walkenbach[12].

The paper introduces a chart of the heat-map type, suitable for several hundred cases and about 25 categorical variables. As discussed in Sec. 4, the design principles are similar to the Caseview approach [9, 10]. The chart is implemented in Excel using standard formulas, sorting and a short VBA macro. The data are stored and processed on one worksheet, and the chart is produced on a separate worksheet. The presentation form of the chart is produced with Excel's page setup feature by specifying that the document printout should be one page long and one page wide. The chart can be exported for publishing or further processing either by printing it to an Adobe[®] PostScript[®] printer assigned to a file, or by using one of the solutions for producing Adobe[®] PDF documents. The charting technique is demonstrated with two studies: one from biomedical informatics, and one from applied biostatistics.

2 Method

The motivating example (Fig. 1 in Sect. 3) depicts accuracy of automated assignment of MeSH[®] thesaurus[13] descriptor headings (variables) to 309 abstracts of medical articles (cases). First, within each abstract, the predicted support for each heading from the automated classification (in the presented example it was a k-nearest neighbour algorithm) was ranked. Then, for each heading actually assigned to a given abstract by a human information specialist (marked by a black cell, which is easily accomplished with conditional formatting in Excel), support rank was depicted in the adjacent cell with dark green colour for high support and dark red colour for low support. Conversely, an actually absent descriptor was depicted with a white cell, and for the adjacent cell the same nine-point two-colour scale for used with the reverse role of high and low support rank. In this way, each case was depicted by one row of a table and each attribute is depicted with two adjacent columns (actual presence and prediction accuracy). For cell colouring, which exceeded the capabilities of conditional formatting, a short macro was written based on the cell font and cell interior ColorIndex property.

Next, a rank-based classification accuracy measure was calculated for each case, and cases in the plot were sorted from better accuracy to worse downwards, thus producing increasing share of red from top to bottom of the chart. Finally, an analogous measure was calculated for each variable by summing discrepancies for a column rather than for a row, and the variables were sorted on the basis of that criterion from left (more accurately predicted attributes, characterised by prevailingly green cells) to right (less accurately predicted attributes, by prevailingly red cells).

The second example (Fig. 2 in Sect. 3) is from clinical biostatistics. The data come from a study on morbidity after conization with 1235 women [11], and the chart depicts classification accuracy for six dichotomous outcome variables. Details are given below.

3   Results

The presented charts primarily serve as demonstrations of the underlying principle, but even with the selected number of cases and variables they provide potentially useful information for the researcher.

In the first example (Fig. 1), the increasing share of red-coloured cells from top to bottom and from left to right shows for which abstracts the algorithm performed better and for which the performance was worse, as well as which descriptor headings were more accurately predicted and which were predicted less accurately. Although in general, the descriptor headings that were more frequently assigned by the human specialist tended to be somewhat less accurately predicted, it is encouraging for practical application to note that the heading labelled H, which was by far the most frequently assigned, lies in the middle in terms of prediction accuracy.

In the second example, the six outcomes – early post-operative presence and prolonged (i.e., more than two months) duration of pain, flour and bleeding, respectively – were predicted with logistic regression from conization technique (cold-knife with modified Sturmdorf stitch, cold-knife with electrocoagulation of conus bed, or large-loop excision) and patient characteristics (age, pre-operative flour, cervical smear result, and post-operative treatment with antibiotics). In the chart (Fig. 2), the 878 cases with no missing outcome are depicted, and cell colour codes classification accuracy based on the 0.5 cut-off value for prediction of outcome presence vs. absence. The outcome variables are sorted on substantial basis: for each of the three symptoms, the short-term outcome is followed by the long-term outcome. The chart clearly shows that early pain, which was frequent, is the least accurately predicted outcome, while the seldom present prolonged bleeding is predicted more accurately, whereby even for the women actually suffering from it, the predicted probability was relatively low (red cells adjacent to the black cells), so a lower cut-off value would be required if the study were oriented towards clinical prediction rather than statistical inference.

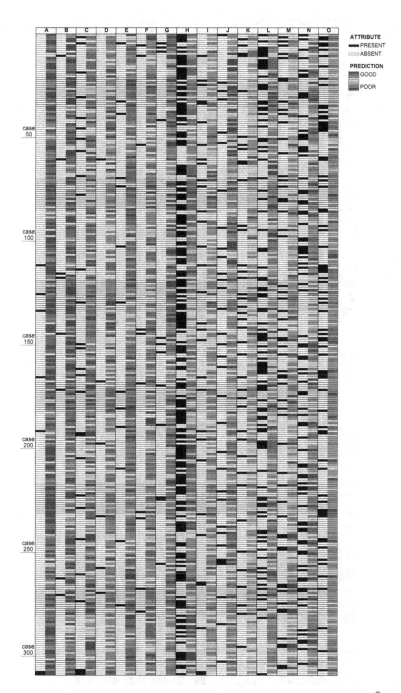

Fig. 1. Spreadsheet cell-chart depicting accuracy of automated assignment of MeSH® thesaurus descriptor headings to abstracts of medical articles (see Sect. 2 for details)

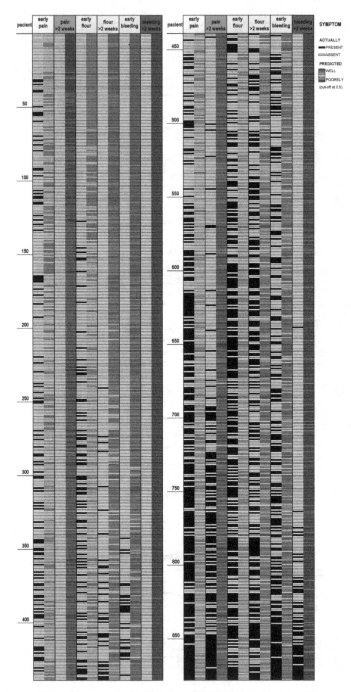

Fig. 2. Spreadsheet cell-chart depicting accuracy of prediction of various post-operative symptoms in a study of morbidity after conization (see Sect. 3 for details)

4 Discussion

The presented charting procedure for visualising up to about a thousand cases and about twenty-five categorical variables with spreadsheet technology has proven to be sucessful. It is useful for comparison of various variables (different outcomes, different classification methods, etc.), and regarding observed values it can easily be extended from binary to polytomous variables.

Our solution is Microsoft® Office based, but it is portable to other platforms, e.g., via implementation in the OpenOffice suite[14]. With the forthcoming enhancements in Excel version 12[15], which will break the 256-column limit and offer unlimited conditional formatting with readily implemented colour scales and symbol coding, the described and similar cell-charts will not require macro programming, and they will offer even more possibilities for data visualisation.

In the presented charts, sorting of cases is based on classification accuracy, but if the visualisation were to serve some other purpose within the research scope, the sorting criterion could be optimised accordingly. For example, for comparison between cases with present and absent outcome, cases could be sorted on the basis of predicted outcome within each variable, thus yielding a block of white cells (absent outcome, i.e., $Y=0$ in terms of logistic regression) followed by a block of black cells (present outcome, i.e., $Y=1$). If retaining the principle of sorting the variables on the basis of prediction accuracy, the resulting chart would be a combination of the presented chart-type with the principle of permutation matrices.

Further applications can easily be envisioned. For example, we successfully applied the procedure in the field of bioinformatics, on a dataset from a genetic study of 382 infertile men [12] that employed data mining methods on a published database of Y-chromosome deletions for the purpose male infertility diagnosis. We selected the five most informative genetic markers out of the total 177, the binary outcome was type of infertility (oligo-asteno-terato-azoospermia or azoospermia), and the chart compared the results of three regression models (categorical regression with optimal scaling with missing predictor value imputation, logistic regression, and probit regression). Cell colouring was based on absolute value of standardised regression residuals, while case sorting was based on the first regression method, since for logistic and probit regression the missing predictor values were not imputed (so the residuals were coloured grey for such cases). Instead of variable sorting, the Phi coefficient of nominal association and area under ROC were listed in the header row as statistically sound measures of prediction accuracy for each method. Without going into details, we can state that the chart provided useful information for the physicians involved in the study.

Applicability to diverse fields of biomedical informatics and similar design principles relate our method to the (generalised) Caseview approach [9, 10]. Namely, like the authors of Caseview, we apply an ordinal colour scale to the cells after having defined a reference frame with three criteria: one nominal variable that can be meaningfully ordered (subject heading in the first example, symptom in the second example), one ordinal dimension (accuracy of case classification), and one binary

---

[14] http://www.openoffice.org
[15] http://blogs.msdn.com/excel/default.aspx

attribute (in our procedure, it is the observed value, which is depicted to the left of prediction accuracy for each attribute).

In general, spreadsheets are very useful for displaying added information within visualisations in the sense of upgrading charts to illustrations, or, in the words of Wilkinson's [13] paradigm, turning a graph into a graphics. When carefully constructed, spreadsheet-based charts from the pixelisation family can follow the principles of good visualisation [14]. As a final point, through changing of model or visualisation parameters with live chart updating, spreadsheets can be an excellent tool for implementing the exploratory $graph \rightarrow fit \rightarrow graph \rightarrow \ldots$ paradigm.

References

1. Friendly, M.: *Visualizing categorical data.* SAS Institute, Cary, NC (2000).
2. Bertin, J.: *Graphics and graphic information-processing.* de Gruyter, New York (1981).
3. Hartigan, J.A., Kleiner, B.: A mosaic of television ratings. *The American Statistician,* 38(1) 32–35, 1984.
4. Friendly, M.: Mosaic displays for multi-way contingency tables. *Journal of the American Statistical Association,* 89(425) 190–200, 1994.
5. Eisen, M.B., Spellman, P.T., Brown, P.O., Botstein, D.: Cluster analysis and display of genome-wide expression patterns. *Proceedings of the National Academy of Sciences of the United States of America,* 8(95) 14863–14868, 1998.
6. Pavlidis, P., Noble, W.S.: Matrix2png: a utility for visualizing matrix data. *Bioinformatics,* 19(2) 295–296, 2003.
7. Heiser, D.A.: *Microsoft Excel 2000 and 2003 faults, problems, workarounds and fixes.* http://www.daheiser.info/excel/frontpage.html
8. Neuwirth, E., Arganbright, D.: *The active modeler – mathematical modeling with Microsoft Excel.* Brooks/Cole, Belmont, CA (2004).
9. Lévy, P.P.: The case view, a generic method of visualization of the case mix. *International Journal of Medical Informatics,* 73(9-10) 713–718, 2004.
10. Lévy, P.P., Duché, L., Darago, L., Dorléans, Y., Toubiana, L., Vibert, J.-F., Flahault, A.: ICPCview: visualizing the International Classification of Primary Care. In: Engelbrecht, R., et al. (eds.): *Connecting Medical Informatics and Bio-Informatics. Proceedings of MIE2005.* IOS Press, Amsterdam (2005) 623–628.
11. Zupancic Pridgar, A.: *The influence of vaginal flora on morbidity after conization* (MSc thesis). University of Ljubljana, Faculty of Medicine, Ljubljana (2003).
12. Džeroski, S., Hristovski, D., Peterlin, B.: Using data mining and OLAP to discover patterns in a database of patients with Y-chromosome deletions. *Journal of the American Medical Informatics Association,* 7(Suppl) 215–219, 2000.
13. Wilkinson, L.: *The grammar of graphics.* Springer, New York, NY (1999).
14. Tufte, E.: *The visual display of quantitative information* (16th printing). Graphics Press, Chesire, CT (1998).

Dynamic Display of Turnaround Time
Via Interactive 2D Images

Peter Gershkovich[1] and Alexander Tselovalnikov[2]

[1] Yale University Medical School Department of Pathology,
New Haven, Connecticut, USA
[2] Signal Studio, Barnaul, Russia

Abstract. One of the key characteristics of the clinical practice of anatomic pathology is turnaround time. However, standard methods of reporting and data representation provide limited insight. Many factors have to be taken into consideration. Significance of various points on a timescale varies depending on days of the week and holidays. The number of pathologists, subspecialty, and case mix– all has to be in front of an observer to understand events and their circumstances. Currently, reporting requires printing of numerous tabular data sets that provide different views of the same event. Through the use of Java 2D API, we demonstrate how 2D graphic, color gradient and interactive display of data delivered via the Web could facilitate analysis of a particular problem of turnaround time in anatomic pathology operations and similar complex events.

1 Introduction

Turnaround time is one of the most critical indicators used for quality assurance and monitoring day-to-day operations in practice of anatomic pathology [1,2,3]. It has significant impact on the satisfaction of clinicians, who rely on timely pathology reports. Yet, it is not obvious how to display turnaround time to provide an immediate insight into the state of operations, to allow timely adjustments to workflow and to guide management decisions that could keep this indicator in check.

One typical approach is to run standard reports via existing electronic systems. These reports provide structurally defined tabular data. Questions that are usually triggered by these reports require additional data. Since such data is not immediately available, an extra effort is required to produce more reports. This cycle seems to be perpetual and costly to sustain. In addition, the volume of such reports is rather large and detailed analysis requires time. Use of such reports could weaken, as they become less of a management tool and more of a ritual.

Another approach is to aggregate data, reducing the amount of detail. Again, significant training would be required to produce and interpret such reports in a meaningful way. For example, evaluating central tendencies would not be very productive because only a few critical cases that miss acceptable threshold could cause an ordering physician to switch to a different lab.

This particular problem is very indicative of the challenge in analyzing temporal data so that a user will get a clear, dynamic, overall view and will be instantly able to access details. Issues of temporal data analysis long have been a focus of interest

P.P. Lévy et al. (Eds.): VIEW 2006, LNCS 4370, pp. 55–62, 2007.

among many groups across various domains from temporal pattern mining algorithms to the art of data visualization.

Many techniques of time series visualization have been well documented [4]. Attempts also have been made to define guidelines for a self-evident display. But even with interesting underlying data, subtle artistry and common sense, these attempts cannot overcome single display limitations – inability to manipulate data and see hidden details. No matter how impressively clear the display is, the observer could make a wrong decision, if even a small part of critical information is missing. Especially dazzling displays, sometimes, create a disservice to the viewer as their superb presentation and elegance substitute for completeness.

On the other hand, complex data mining techniques are expensive to implement and such tools have a long way to become widely available. Humans, however, may effectively perform visual analysis especially when dealing with identification of patterns [5] that are based on common knowledge not necessarily available to the system. In particular, macro events such as the start of a flu season or the departure of a key employee may be well known to an observer but are not recorded in existing systems.

In this work, we suggest that existing technology allows us to create a framework for efficient data visualization that would provide additional analytical capabilities in evaluating turnaround time in pathology practice. We also suggest that the framework described here can be used in other areas with similar data visualization requirements.

2 Implementation and Discussion

One of the principles in the selection of technology was its adherence to open standards. Our goal, in that regard, was to reduce the cost of ownership, facilitate knowledge transfer and minimize maintenance of the system. Another hope was to reduce client machine requirements from both performance and software installation standpoints. The only requirement to the client machine we could not avoid was the presence of the World Wide Web Consortium (W3C) standards compliant browser. In particular, such browsers should be able to handle JavaScript (a language commonly used for the client side user interface manipulation), provide Document Object Model (DOM) access and have Cascading Style Sheets (CSS) support. We tested Safari and FireFox web browsers on Mac OS X. FireFox also worked fine on Linux and Windows operating systems. Relatively fast network communication was also assumed. Typical DSL or Cable connections were deemed sufficient to provide a satisfying experience. Internal TCP/IP network was used to connect computers with server class machines.

The server side had to handle data storage, retrieval, preprocessing and rendering of images. This way, we could eliminate client-side issues completely and preserve control over required computational power and virtual memory.

We selected Java as our programming language of choice. Systems designed with Java could run on a variety of platforms and, if designed well, allow easy maintenance and extensibility.

Our particular implementation was based on Mac XServe running Java 1.5. The underlying data has been stored in a relational format in MySql database. Figure 1 shows a high level of the system's architecture.

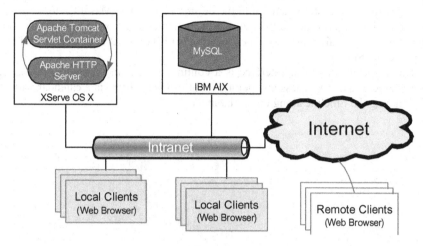

Fig. 1. System architecture

Apache HTTP server, along with Apache Tomcat Servlet Container, has been used to render HTTP requests. These two components resided on the same server connected via mod_jk connector. The application itself has been designed using Model View Controller (MVC) paradigm well described in literature [6,7]. The images were rendered via Java 2D API. One of the reasons of going with Java 1.5 was improved image rendering. See documentation for a complete list of new features [8].

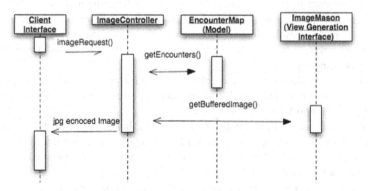

Fig. 2. Simplified UML sequence diagram

Application design can be better understood via a Unified Modeling Language (UML) sequence diagram shown in Figure 2.

The user interaction with the browser interface causes, besides typical events, asynchronous requests to the server. The ImageController (a subclass of an abstract Controller) determines what part of the data model to use and gets a required map of cases or encounters. When all required data is collected the Controller sends a request to the ImageMason interface to build the desired image. The MasonFactory class (not shown in the diagram) is used to get one of its implementations based on the type of client request. Finally, an image is encoded into JPEG format to be understood by the requester's browser.

The key element of our interface is a continuous timeline shown in Figure 3. A user is able to quickly access any point on that timeline ether via a calendar widget or by dragging the timeline along the horizontal axis.

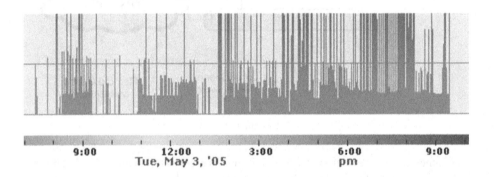

Fig. 3. Continuous timeline with daily turnaround reports

Visual aids of color and color gradient indicate transition between day and night as well as weekends and holidays. This way, the user can move from one day to another rapidly relating data to the time of actual event. The timeline and associated reports are rendered dynamically inside the mobile layer creating a smooth transition effect. That effect is created by loading images behind the scene via asynchronous calls to the server; otherwise, users would see a typical delay while each individual block of the display is being loaded.

It is important to notice that by creating a smooth transition between reports, we eliminate a problem that traditional linking has – the presence of an arbitrary defined split between two reports.

The report in Figure 3 shows individual cases that had to be evaluated by pathologists. The vertical axis shows a sequence of pixel-wide bars representing the time it took to process each individual specimen as a percentage of established thresholds. For example, some surgical biopsies had an established threshold of 48 hours vs. complex cases requiring special procedures that could take, normally, a week. The black horizontal line that crosses the middle of the report area is the threshold. The gray pixel-wide bars that reach that line represent cases that stayed below threshold in respect to the time it took to process a case. Bars, displaying cases that took more time to process, are rapidly changing color to red as they cross the established benchmark.

Such visualization can be viewed as a variation of pixel-oriented techniques [9,10] where each pixel on the diagram corresponds to the time it took to process each individual case. The color of each pixel does not change if they are located below threshold. If, however, they are located above threshold we introduce an attribute - significance of the delay. This attribute is expressed through the changing color gradient. We limited the height of bars so they would not exceed the two hundred percent mark. Such design was, partially, necessary to overcome a problem with bars in general as they have a tendency to take a large space [10]. It was a creative approach that does not have an underlying formalism. However, from the analysis standpoint, showing cases that took twice more time to process than established threshold were not as critical for review on that level of detail.

The goal of continuously displaying various performance patterns was achieved at this point. These patterns are immediately recognizable. For example, a pattern in Figure 3 clearly demonstrates that cases accessioned in the afternoon hours are more likely to be delayed. An even more dramatic difference can be seen on cases accessioned on Fridays (see Figure 4). A large number of them take too long to process.

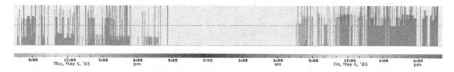

Fig. 4. A pattern shows significant change in turnaround time on Friday

It is clear, even without going into further details, that improving staff coverage in the afternoon and on the weekend would have the most effect on performance. Applying various filters, such as signing physician or type of specimen, could further clarify our understanding of underlying process.

Filtering can go only so far as the resolution goes. In order for us to see more detail we created zoom-in capabilities by adding a clickable map behind each report. This map allows us to zoom into each hour. After double-clicking the report area similar to the one shown in Figure 4, a more detailed report is displayed as shown in Figure 5. Here, our threshold is displayed as a thick red line. The purpose of that line is to split all cases in three categories: cases with below-threshold turnaround, cases that were above threshold but within reasonable range, and cases shown in red that were significantly delayed.

Such division, again, allows a viewer to make a decision to seek improvement in extreme cases first. What may be less obvious, however important, is the fact that some cases were processed significantly faster than turnaround threshold. Further analysis of these cases may provide additional insight on how to organize workflow and to avoid extremes.

Finally, a red arrow below the graph can be moved along the timeline so that a group of actual cases will be shown in a detailed tabular form as shown below.

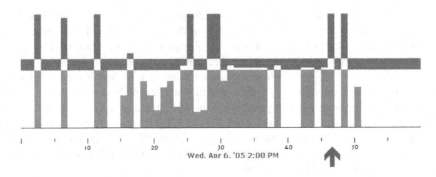

Fig. 5. One-hour report separates all cases in three categories. A movable red arrow points to a segment of a graph that is displayed in tabular form with maximum detail.

These cases, again, are displayed dynamically covering a 10-minute range. If required, one could study turnaround time on that level to discuss each case individually. Just a small step is required to open up a case in existing clinical system and review all available details.

Specimen Type	Accessioned	Signed	Actual Turnaround	Threshold
CN	04/06/2005 12:28	04/07/2005 16:14	27 hrs	24 hrs
SR	04/06/2005 12:29	04/07/2005 11:38	23 hrs	48 hrs
CN	04/06/2005 12:30	04/07/2005 16:14	27 hrs	24 hrs
BPS	04/06/2005 12:31	04/08/2005 08:35	44 hrs	72 hrs
SR	04/06/2005 12:31	04/08/2005 14:19	49 hrs	48 hrs
BPS	04/06/2005 12:35	04/07/2005 12:55	24 hrs	72 hrs
SR	04/06/2005 12:35	04/11/2005 20:30	5 days	72 hrs

Fig. 6. Detailed tabular report covering 10 minutes interval around selection point

There is no doubt that visualization of data is a viable analytical tool. Many methods of data visualization have been proven to provide a better assistance in the analysis of complex events. Such methods could potentially improve the performance and decision-making process in the practice of anatomic pathology. In addition, such techniques of data manipulation and data interaction as brushing, filtering, zooming and linking, described and prototyped in many systems [11, 12], can further assist in making visual analysis more effective. Pixel-oriented techniques for transforming data into visual form proved to be successful in displaying large volumes of data [10]. Significant increase in computing power and the speed of electronic communications created good underlying infrastructure to use visualization techniques for a broader set of users. Open standards and Open Source tools make development and cost of ownership for such systems feasible for practical implementation.

We identified a set of requirements to build a system that would feed well into practice of anatomic pathology and made an emphasis on human ability to identify and discover patterns. Our initial display of turnaround time data was built to utilize that capability. We suggest that it may be more effective than other forms of analysis provided we would use data interaction techniques to get more details when necessary. Pathologists, who are the main users of our system may benefit even more as they are in general more facile at assimilating visual information vs. information presented in a tabular form.

Our framework attempts to create a data flow parallel to existing workflow. Potentially, it could provide smooth transition into just-in-time analysis, do prognostic reporting and allow users to see projections. This framework can be used in similar situations when time-series monitoring is necessary.

3 Conclusion

We were able to demonstrate the feasibility of displaying turnaround time data using a visual model. Using advances in current technology and in visualization techniques, it is becoming possible to deliver more data as well as better visual analysis tools to a broader audience. However, the impact on productivity and performance that such visualization may have remains unclear. Further research is required to compare effectiveness of visual reporting tools vs. existing reports on quality assurance and outcomes. On a higher level it is also necessary to create collaboration ties between scientists representing all domains with stakes in visual analysis as well as graphic designers and systems developers. Such interdisciplinary collaboration may help to create a solid foundation for practical and theoretical advances in data visualization and visual analysis techniques.

Acknowledgements

The authors would like to thank John H. Sinard, M.D., Ph.D., Director of Program for Pathology Informatics at Yale, for critically reading this paper and for his valuable advice.

References

1. Novis DA, Zarbo RJ, Saladino AJ. Interinstitutional comparison of surgical biopsy diagnosis turnaround time: a College of American Pathologists Q-Probes study of 5384 surgical biopsies in 157 small hospitals. Arch Pathol Lab Med. 1998 Nov; 122(11):951-6.
2. Galloway M, Nadin L. Benchmarking and the laboratory. J Clin Pathol. 2001 Aug; 54 (8):590-7.
3. Novis DA, Walsh MK, Dale JC, Howanitz PJ; College of American Pathologists Q-Tracks. Continuous monitoring of stat and routine outlier turnaround times: two College of American Pathologists Q-Tracks monitors in 291 hospitals. Arch Pathol Lab Med. 2004 Jun; 128 (6):621-6.
4. Edward R. Tufte, The visual display of quantitative information, Graphics Press, Cheshire, CT, 1986
5. Payne, D. G. and Wenger, M. J. (1998) Cognitive Psychology. Boston: Houghton Mifflin.
6. Buschmann, F., Meunier, R., Rohnert, H., Sommerlad, P., & Stal, M. (1996) Pattern-oriented software architecture: A system of patterns. Chichester, UK: John Wiley & Sons.

7. Erich Gamma, Richard Helm, Ralph Johnson, John Vlissides. (1994) Design Patterns: Elements of Reusable Object-Oriented Software. Addison-Wesley Professional Computing Series, Addison-Wesley, Reading Mass.

8. http://java.sun.com/j2se/1.5.0/docs/guide/2d/new_features.html

9. Keim, D. A. 2000. Designing Pixel-Oriented Visualization Techniques: Theory and Applications. IEEE Transactions on Visualization and Computer Graphics 6, 1 (Jan. 2000), 59-78.

10. Keim, D. A., Hao, M. C., Dayal, U., and Hsu, M. 2002. Pixel bar charts: a visualization technique for very large multi-attribute data sets. Information Visualization 1, 1 (Mar. 2002), 20-34.

11. Wong, P. C. and Bergeron, R. D. 1997. Brushing techniques for exploring volume datasets. In Proceedings of the 8th Conference on Visualization '97 (Phoenix, Arizona, United States, October 18 - 24, 1997). R. Yagel and H. Hagen, Eds. IEEE Visualization. IEEE Computer Society Press, Los Alamitos, CA, 429-ff.

12. Buja, A. McDonald, J.A. Michalak, J. Stuetzle, W. Interactive data visualization using focusing and linking Visualization, 1991, Proceedings, IEEE Conference 156 - 163

Pixelizing Data Cubes: A Block-Based Approach

Yeow Wei Choong[1,2], Anne Laurent[3], and Dominique Laurent[2]

[1] HELP University College, BZ-2 Pusat Bandar Damansara, 50490 Kuala Lumpur, Malaysia
choongyw@help.edu.my
[2] ETIS-CNRS, Université de Cergy Pontoise, 2 av. Chauvin, 95302 Cergy-Pontoise, France
dominique.laurent@dept-info.u-cergy.fr
[3] LIRMM-CNRS, Université Montpellier 2, 161 rue Ada, 34392 Montpellier, France
anne.laurent@lirmm.fr

Abstract. Multidimensional databases are commonly used for decision making in the context of data warehouses. Considering the multidimensional model, data are presented as hypercubes organized according to several dimensions. However, in general, hypercubes have more than three dimensions and contain a huge amount of data, and so cannot be easily visualized. In this paper, we show that data cubes can be visualized as images by building blocks that contain *mostly* the same value. Blocks are built up using an APriori-like algorithm and each block is considered as a set of pixels which colors depend on the corresponding value. The key point of our approach is to set how to display a given block according to its corresponding value while taking into account that blocks may overlap. In this paper, we address this issue based on the *Pixelization paradigm*.

1 Introduction

On-Line Analytical Processing (OLAP) is currently a major research area in the database community aiming at managing huge amounts of data for decision making. The modelling of a multidimensional database has been extensively studied [28] and data visualization is one of the major issues in this context [20,17,18].

In this paper, we consider a complete process of multidimensional databases visualization. This process relies on the discovery of blocks of homogeneous data which are then displayed to the user. The complete process can thus be considered as a two-layer process, as in [20]:

- a logical layer that deals with the discovery of blocks of similar measure values,
- a presentational layer that deals with the presentation of the blocks to the user.

In this paper, we give a brief overview of the techniques used in discovering blocks and we focus on the presentational layer.

The rest of the paper is organized as follows. We survey the concepts of data warehouse and OLAP in Section 2. Section 3 provides the background for block discovery and visual data analysis is introduced in Section 4. Finally, Section 5 concludes our work and presents topics for future work.

P.P. Lévy et al. (Eds.): VIEW 2006, LNCS 4370, pp. 63–76, 2007.

2   Data Warehouses

2.1   Main Features

A primary objective of a data warehouse is to provide intuitive access to information stored in a database that is used in decision making. According to [14], "A data warehouse is a subject-oriented, integrated, non-volatile and time-variant collection of data in support of management's decision making process".

Data warehousing refers to the process of constructing and exploitating of the data warehouse. Data warehousing thus includes the integration of the data from multiple sources into a unified schema at a single location to facilitate data analysis for decision making. Thus, the construction of a data warehouse includes data integration, data cleansing, data consolidation and On-Line Analytical Processing (OLAP) [5,9,12].

OLAP systems are constructed in a data warehouse environment that serves as a repository of the data to be processed. From a data warehouse perspective, data mining can be viewed as an advanced stage of on-line analytical processing [12].

2.2   Multidimensional Data Model

The most popular data model for a data warehouse is a multidimensional data model (MDD). MDD models typically have two types of data: (a) *measures* which are numerical in nature and (b) *dimensions* which are mostly textual data characterizing the measures [24].

A typical MDD example is in the case of a retail business where *Product*, *Location* and *Time* are dimensions and *Quantity*, *Price* and *Sales* are measures. Intuitively, it is possible to think of a business as a data cube with these dimensions and measures, as shown in Figure 1. Any point, known as *cell* in OLAP literature, is characterized by a value on each dimension and by the corresponding measure values. Thus, one can think of a cell in the cube as the measurements of the business (e.g. *Quantity*, *Price* and *Sales*) for a particular combination of *Product*, *Location* and *Time*.

Given the above structure, OLAP tools provide fast answers to the ad-hoc queries that aggregate the data from the data warehouse. Such process is known as multidimensional data analysis or OLAP analysis.

Generally, a multidimensional database is a set of *hypercubes* (hereafter *cubes*). We note that *hierarchies* may be defined over dimensions, for instance to describe sales in function of states and not of cities. As hierarchies are not considered in the present approach, we shall not consider this feature in the following definition.

Definition 1 – Cube. *A k-dimensional cube, or simply a cube, is a tuple $\langle C, dom_1, \ldots, dom_k, dom_m, m_C \rangle$ where*

- *C is the name of the cube,*
- *$dom_1, \ldots, dom_k$ are k finite sets of symbols for the members associated with dimensions $1, \ldots, k$ respectively,*
- *$dom_m = dom_{mes} \cup \{\bot\}$, dom_{mes} is a finite totally ordered set of all possible measure and $\bot$ is a constant not in dom_{mes} that represents null values.*
- *m_C is a mapping from $dom_1 \times \ldots \times dom_k$ to dom_m.*

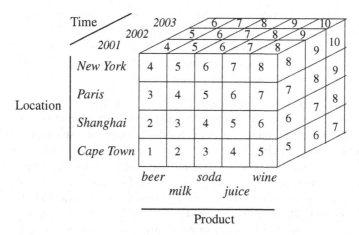

Fig. 1. A three-dimensional data cube

A cell c of a k-dimensional cube C is a $(k+1)$-tuple $\langle v_1, \ldots, v_k, m \rangle$ such that, for every $i = 1, \ldots, k$, v_i is in dom_i and $m = m_C(v_1, \ldots, v_k)$. Moreover, m is called the *content* of c and c is called an *m-cell*.

As pointed out in [7], a cube can be associated with different *representations*, depending on the way member values are ordered in the sets dom_i ($i = 1, \ldots, k$).

Definition 2 – Representation. *A representation of a cube C is a set $R = \{rep_1, \ldots, rep_k\}$ where for every $i = 1, \ldots, k$, rep_i is a one-to-one mapping from dom_i to $\{1, \ldots, |dom_i|\}$.*

Figure 2 displays two different representations of the same cube. In this paper, we consider a *fixed* k-dimensional cube C and a fixed representation of C, $R = \{rep_1, \ldots, rep_k\}$.

PRODUCT

	C1	C2	C3	C4	C5	C6
P1	6	6	8	5	5	2
P2	6	8	5	5	6	75
P3	8	5	5	2	2	8
P4	8	8	8	2	2	2

C1 C2 C3 C4 C5 C6 **CITY**

(a)

PRODUCT

	C3	C1	C2	C4	C5	C6
P4	8	8	8	2	2	2
P2	5	6	8	5	6	75
P1	8	6	6	5	5	2
P3	5	8	5	2	2	8

C3 C1 C2 C4 C5 C6 **CITY**

(b)

Fig. 2. Two representations of a data cube

2.3 OLAP Operations

As mentioned previously, in the multidimensional data model, data are organized according to multiple dimensions, and each dimension contains multiple levels of abstraction defined by concept hierarchies. This way of representing the data allows the user to view the data from different levels of detail. To achieve this, a number of OLAP operations are available for navigating through the data set. Navigation is a query-driven process, and a number of proposals have investigated formal models and languages to this end (see [12,22,29] for surveys). Some typical OLAP operations on multidimensional data are the following ones:

- *Roll-up.* This operation performs aggregation on a data cube. This is done by either *climbing up* a concept hierarchy for a dimension or by *dimension reduction.* Essentially, this operation increases the level of abstraction. An example of a roll-up operation is aggregating the *sales* of all cities to the level of country.
- *Drill-down.* This operation is essentially the reverse of roll-up. It puts more detail to a given dimension, thus decreasing the level of abstraction. For example, this operation *steps down* a concept hierarchy from the level of *month* to the level *day*.
- *Slice and dice.* The *slice* operation performs a selection on one dimension of a given cube, resulting in a subcube. For example, a selection of the months April, May and June from the *month* dimension is a *slice* operation. The *dice* operation defines a subcube by performing a selection on two or more dimensions.
- *Pivot* or *rotate.* This operation is a visualization operation that rotates the axes of the data in order to provide a different presentation of the data.
- *Switch.* This operation performs the interchanging of the positions of two members of a dimension of a cube.

Each of the above operations is illustrated in Figure 3, where the original cube is taken from Figure 1. There are many other OLAP operations such as *push*, *pull*, *join* and *merge* that are not covered here. A detailed list of OLAP operators and their descriptions can be found in [22]. More details on data warehouse and OLAP can be found in [3,10,11,14,15,19].

2.4 Data Warehouse Visualization

For effective data analysis, it is important to include the user in the data exploration process and to combine the flexibility, creativity and the common sense knowledge of human with the computational power of the computers [4]. The fundamental idea of visual data analysis is to present the data in some visual forms, allowing the user to gain insight into the data [25]. Data visualization involves examining data represented by dynamic images (colors) rather than pure numbers. Normally these images are rendered in colors rather than shades of gray, that can be used as data value indicators [21].

Visual data exploration usually follows a three-step process (*overview first*, *zoom and filter*, and *details-on-demand*) known as the Information Seeking Mantra [23].

The process of visualization can be considered as a kind of numerical data transformation. In the context of OLAP, this transformation process is performed on measure values. In the last decade, a large number of novel information visualization techniques have been developed, allowing visualizations of multidimensional data sets [13], and supporting the hierarchical structure of the dimensions of a data cube [26].

In [16], a presentational model for OLAP data is introduced. This model is based on the geometrical representation of a cube and the human perception in the space, and it is shown how the logical and the presentation models can be integrated.

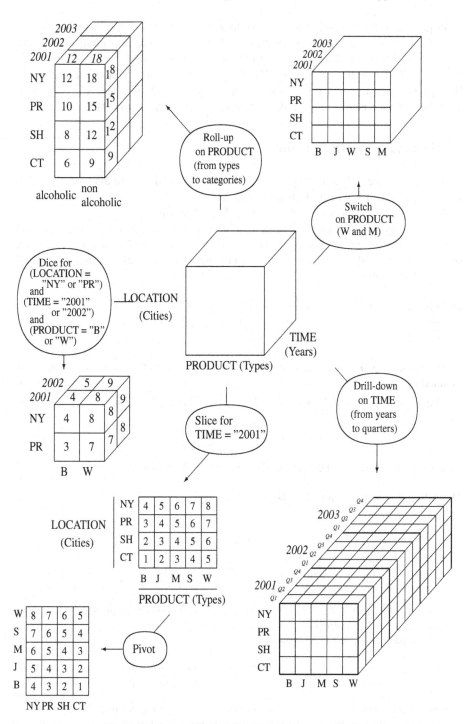

Fig. 3. Examples of typical OLAP operations on a cube

In [2], the authors present an approach to visualizing multidimensional data in the presence of constraints and user preferences. In this approach, constraints refer to the fact that the device on which data are displayed imposes restrictions (e.g. the size of the screen), and user preferences are seen as partial orderings over the member values. In this approach, dimensions are nested so as to fit the constraints, whereas in this paper, we focuss on the visualization of blocks that summarize large data cubes.

3   Summarizing Data Cubes

In this section, we review from [8] the APriori-like method used to discover blocks of homogeneous data from a multidimensional database.

Given a k-dimensional cube C and one of its representations $\{rep_1, \ldots, rep_k\}$, let d_i be a dimension of C. Two elements v_1 and v_2 in dom_i are said to be *contiguous* if $|rep_i(v_1) - rep_i(v_2)| = 1$. Moreover, for all v and v' in dom_i, the *interval* $[v, v']$ is the set of all contiguous values between v and v'. In our approach, a block of C is a subcube of C defined as follows.

Definition 3 – Block. *Given a k-dimensional cube C, a block b is a set of cells in C defined by $b = \delta_1 \times \ldots \times \delta_k$ where δ_i is an interval of contiguous values from dom_i, for $i = 1, \ldots, k$.*

Note that in the case where the interval on dimension d_i is the whole set dom_i, δ_i is denoted by *ALL*.

The *relative size* of a block b in C, denoted by $size(b)$, is the number of cells in b, i.e.,

$$r\text{-}size(b) = \frac{\#\,cells\ in\ b}{\#\,cells\ in\ C}$$

Moreover, the support and confidence of a block are defined as follows.

Definition 4 – Support and Confidence. *Let C be a k-dimensional, b a block in C and m a measure value.*

1. *The support of b for m, denoted by supp(b, m) is the ratio*

$$supp(b, m) = \frac{\#\,cells\ in\ b\ containing\ m}{\#\,cells\ in\ C}$$

   Considering a user-given minimum support threshold σ and a measure value m, a block b such that $supp(b, m) > \sigma$ is called σ-frequent for m.
2. *The confidence of b for m, denoted by conf(b, m), is defined as:*

$$conf(b, m) = \frac{\#\,cells\ in\ b\ containing\ m}{\#\,cells\ in\ b}$$

Given a k-dimensional cube C, a support threshold σ and a confidence threshold γ, we have presented in [8] an algorithm for computing a set $\mathcal{B}$ of blocks of C such that for every b in $\mathcal{B}$, there exists a measure value m in C for which $supp(b, m) > \sigma$ and

$conf(b,m) > \gamma$. As argued in [8], the blocks in $\mathcal{B}$ are meant to summarize the content of the cube C, either by associating them with a rule (that can be fuzzy due to overlappings) or by visualizing them appropriately, which is the subject of the present paper. Roughly speaking, our algorithm works as follows:

1. For every $i = 1, \ldots, k$, compute all maximal intervals I of values in dom_i such that, for every v in I, the block $b_v = \delta_1 \times \ldots \times \delta_k$ where $\delta_i = [v,v]$ and $\delta_j = ALL$ for $j \neq i$, is σ-frequent.
2. Combine the intervals in a level wise manner as follows: at level l, compute all σ-frequent blocks $b = \delta_1 \times \ldots \times \delta_k$ such that exactly l intervals defining b are different than ALL. Assuming that all σ-frequent blocks have been computed at the previous levels, this step can be achieved in much the same way as frequent itemsets are computed in the algorithm Apriori (see [8] for more details).
3. Considering the set of all blocks computed in the previous step, sort out those that are not minimal with respect to the inclusion ordering and then those having a confidence for m less than or equal to γ.

We illustrate the computation of the blocks based on the cube C and its representation as displayed in Figure 2(a). This cube has two dimensions, namely CITY and PRODUCT and contains 24 cells. Moreover, in this example, we consider the following thresholds: $\sigma = 1/15$ and $\gamma = 3/5$.

According to the definitions of support and confidence given previously, this means that, given a measure value m in C, in order to be output, a block b must contain at least 2 m-cells and that, at least 3 of its cells out of 5 are m-cells. Considering the measure value 8, the three steps above are run as follows:

1. For dimension CITY, we compute successively the supports of the blocks $[c,c] \times ALL$, for every c in $\{C1, \ldots, C6\}$. This computation gives the only interval $[C1, C3]$ since
   (i) $C1$, $C2$ and $C3$ are contiguous values such that $supp(C1,8) = supp(C2,8) = supp(C3,8) = 1/12$ and $1/12 > 1/15$,
   (ii) for $c \in \{C4, C5, C6\}$, $supp(c,8) \leq 1/15$.
   Similarly, for dimension PRODUCT, we obtain the interval $[P3, P4]$.
2. Since we have only two dimensions, this step only combines the two intervals above, which gives the block $b = [C1, C3] \times [P3, P4]$. Since $supp(b,8) = 1/6$, this block is $1/15$-frequent and thus is considered at the next step.
3. The confidence of b is computed as the ratio $4/6$ because b contains 6 cells among which 4 are 8-cells. Since this ratio is greater than $3/5$, b is output as the only 8-block.

Similar computations are run for every measure value other than 8 occuring in C, leading to the following blocks shown in Figure 4: $[C1, C1] \times [P1, P2]$ for measure value 6, $[C3, C4] \times [P1, P3]$ for measure value 5, and $[C4, C6] \times [P3, P4]$ for measure value 2.

In [8], one rule per block is automatically built up, and in order to convey block overlappings, these rules are fuzzy. In this paper, we do not consider rules, but rather, we define a policy for the visualization of the blocks, according to criteria defined by the user.

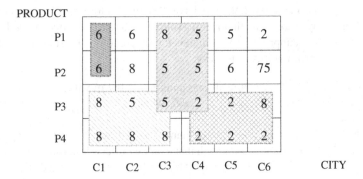

Fig. 4. Data cube and associated blocks

4 Visualization of the Blocks

4.1 Dealing with Multidimensionality

The multidimensionality of the data prevents users from having a global view of the data in a glance, since one cannot visualize more than two or three dimensions at a time. Since only a 2- or 3-dimensional subcube can be visualized, our approach aims to provide a way for displaying the intersections of the blocks with this subcube.

To this end, given a k-dimensional cube C and the set B of associated blocks, we assume that the user can choose

– the 2 (or 3) dimensions that are to be displayed, and
– a measure value for all other dimensions.

In this way, the corresponding 2- or 3-dimensional cube to be displayed is a block b_0 and for each b in B overlapping b_0, the intersection $b \cap b_0$ is also a 2- or 3-dimensional block.

Example 1. *Let us consider a cube containing sale counts according to the following four dimensions: Product, City, Month and Customer-Category. If a user chooses the two dimensions Product and City as being the displayed dimensions, and if this user chooses particular values for the dimensions Month and Customer-Category, say m_0 and c_0, respectively, then the block b_0 is defined by $b_0 = dom_1 \times dom_2 \times [m_0, m_0] \times [c_0, c_0]$. Thus $dom_1 \times dom_2$ can be displayed on the device together with a reference to the values m_0 and c_0.*

Moreover, for every block $b = \delta_1 \times \delta_2 \times \delta_3 \times \delta_4$ in B such that $m_0 \in \delta_3$ and $c_0 \in \delta_4$, the block $\delta_1 \times \delta_2$ can be highlighted as a sub block of $dom_1 \times dom_2$.

In what follows, given a k-dimensional cube, we assume that the 2- or 3-dimensional block b_0 that is to be visualized is known.

4.2   From Blocks to Pixels

Encoding a quantitative variable such as measure value requires only one psychophysical variable such as hue. In the process of encoding a measure domain, a predefined palette is considered to select the color for each domain entry. The colors in the palette are presented in a color spectrum, predominantly on hue [27]. Colors are ordered according to the measure values to display a gradual transformation of colors according to the gradual transformation of the measure values.

As blocks have several properties (e.g. support, confidence, relative size), we consider the HSB (Hue, Saturation, Brightness) encoding in order to convey several chracteristics of the block within a single pixel. HSB encoding is represented as a cone, as shown in Figure 5.

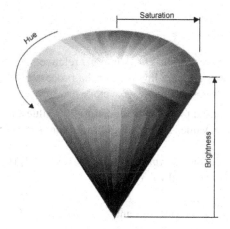

Fig. 5. HSB color space

Let C be the cube to be visualized, M (respectively m) the maximum (respectively minimum) measure value appearing in one block of C, and H_M be the maximum hue value being considered. For each measure value μ, a hue is assigned according to the following formula:

$$Hue(\mu) = \frac{H_M(\mu - m)}{M - m}$$

Note that the hue is expressed as an angle ranging from 0 to 360. In our approach, we map the measure values to an angle ranging from 0 to H_M. We argue that H_M must not be equal to 360 so that the highest measure values do not resemble to lowest ones. For instance, we can consider $H_M = 300$.

For each block associated with a measure value μ, the cells are normally displayed in the hue corresponding to μ. Moreover, as in [1], in order to point out exceptions or outliers, we also consider that cells having a measure value *very different* from μ can be displayed in the hue corresponding to their own value, instead of the hue corresponding to μ. In order to decide whether a cell measure value is *very different* from the block measure value, a threshold expressed as a percentage has to be considered. Denoting by π_H this threshold, μ' is said to be *very different from μ* if

$$\frac{|\mu - \mu'|}{|\mu|} \cdot 100 > \pi_H$$

Saturation and Brightness are also computed regarding the blocks, except for outliers. These values are set to 100% for the outliers, and range from 0 to 100% for any other cell in a block, based on the following criteria:

– the support,
– the relative size,
– the confidence,
– the average measure value,
– the number of outliers.

We note that some of these criteria (*i.e.,* support, relative size, confidence) are very efficiently computed since they are involved in the process of block discovery, while others require additional computations (*i.e.,* average measure value, number of outliers).

Let γ_S be the criterion used to compute the Saturation value and γ_B the criterion used to compute the Brightness value. The set of all blocks can be ordered according to their values for γ_S and for γ_B.

Given a block b, we denote by $rank_S(b)$ (respectively $rank_B(b)$) the rank of b with respect to γ_S (respectively γ_B), and we have:

$$Saturation(b) = \frac{rank_S(b)}{|B|} \cdot 100 \quad \text{and} \quad Brightness(b) = \frac{rank_B(b)}{|B|} \cdot 100$$

where $|B|$ is the number of blocks computed from the cube C. Notice that, if no criterion is defined by the user, then a default value will be used for Saturation and Brightness.

4.3   Managing Overlappings

As seen previously, our approach for computing the blocks implies that some of them may overlap. In that case, one way to display the cells shared by several blocks could be to "mix" the hues associated to these blocks. However, we claim that this could be confusing for the user, because mixing hues may result in another hue corresponding to a different measure value. Instead, we propose to display overlapping blocks in a fore- and background manner, as explained below.

If a given cell is shared by several blocks, we have to decide which block is the most relevant for this cell, so as to determine which block to display on the foreground. To this end, we consider an additional criterion, chosen among those listed above.

Then, this criterion, denoted by γ_F, defines an ordering according to which the blocks are displayed in a fore- or background manner. More precisely, if b and b' are two overlapping blocks such that the value of γ_F for b is greater than that for b', then the overlapping area of b is displayed on the foreground.

Algorithm 1 shows how the blocks are displayed, assuming that the three criteria γ_F, γ_S and γ_B have been chosen by the user. Note that cells belonging to several blocks are treated more than once, which guarantees that they appear with the HSB code corresponding to the block displayed in the foreground based on γ_F.

Data   : C cube (containing measure values ranging from m to M and where m_C is the function associating each cell to its measure value), B set of blocks in C, γ_F foreground criterion (mandatory), γ_S Saturation criterion, γ_B Brightness criterion, π_H outlier threshold, H_M Hue maximal angle

Result   : Pixelization(C)

Order B according to γ_F ;
Order B according to γ_S ;
Order B according to γ_B ;
foreach $b \in B$, *in ascending order according to* γ_F **do**

    Let $m(b)$ be the measure value associated to b ;

    $\text{Hue}(b) \leftarrow \frac{H_M(m(b)-m)}{M-m}$;

    $\text{Saturation}(b) \leftarrow \frac{rank_S(b)}{|B|} \cdot 100$;

    $\text{Brightness}(b) \leftarrow \frac{rank_B(b)}{|B|} \cdot 100$;

    foreach *cell* $c \in b$ **do**

        if $\frac{|m_C(c)-m(b)|}{|m(b)|} \cdot 100 > \pi_H$ **then**

            $\text{Hue}(c) = \frac{H_M(m_C(c)-m)}{M-m}$

        else

            $\text{Hue}(c) = \text{Hue}(b)$

        $\text{Saturation}(c) \leftarrow \text{Saturation}(b)$
        $\text{Brightness}(c) \leftarrow \text{Brightness}(b)$

Algorithm 1. Algorithm for displaying the blocks

The following example illustrates Algorithm 1 above.

Example 2. *Referring back to Example 1, we assume that the blocks to be displayed are those shown in Figure 4. Moreover, we assume that the threshold π_H has been set to 50% and that the criteria associated with γ_F, γ_S and γ_B are respectively the confidence, the average measure value and the number of outliers.*

The table below summarizes the computations of the associated values for each block and Figure 6 displays what users can visualize in this case as output by Algorithm 1. It can be seen in this figure that the cells $(P3, C6, 8)$ and $(P1, C3, 8)$ are computed as outliers, both associated with the same measure value 8. Therefore, their hue, saturation and brightness are respectively 300, 100 and 100. Note that the cell $(P3, C4, 2)$ is computed as an outlier as well for the 5-block. However, this cell also belongs to the 2-block in which it is not an outlier. Since the 5-block is put at the back of the 2-block (because of their respective confidences), the cell $(P3, C4, 2)$ is not displayed as an outlier. On the other hand, the cells $(P3, C2, 5)$ and $(P3, C3, 5)$ are displayed as outliers.

Block	Average	$rank_S(b)$	number of outliers	$rank_B(b)$	Confidence	$rank_F(b)$	H	S	B
2-Block	3	1	1	1	83.33%	2	0	25	50
5-block	5	2	2	3	66.67%	3	150	50	75
6-block	6	3	0	4	100%	1	200	75	25
8-block	7	4	2	2	66.67%	3	300	100	75

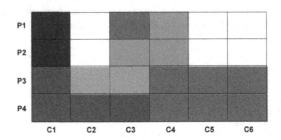

Fig. 6. Visualization example

5   Conclusion

In this paper, we have extended our work on block discovery in OLAP environment to the effective presentation of these blocks using pixelization techniques. This approach is not only particularly relevant to OLAP but it serves as a contribution to the enhancement of multidimensional data analysis.

However, we recall that, in the present approach, we assume that the blocks are available for visualization. We note in this respect that, due to the fact that measure values in a cube are numerical, it is likely that few blocks of similar values can be discovered. This is clearly a limitation for the visualization, that can be overcome by discretizing measure values before computing the blocks. We are currently considering this issue and preliminary results can be found in [6]. Another limitation of our approach is that of being able to display a two-dimensional cube on a given device. Indeed, the number of member values on each displayed dimension is generally very high and so, might not fit on the device. Therefore, constraints have to be taken into account in our visualization technique, and we think that the work reported in [2] provides an appropriate framework to deal with this problem.

A preliminary implementation of the computation of the blocks and their visualization is now available and shows the effectiveness of our approach. We note that the visualization approach proposed in this paper does not require significant additional computation. We are currently working on some extensions, including using other existing methods of HD data visualization, designing non linear scales for the computation of the hue according to the distribution of the measure values, visualizing overlappings more accurately (*e.g.*, using transparency), and pointing out outliers based on additional paradigms (*e.g.*, using different textures or flashing colors).

Based on this work, we plan to study how to extend our visualization technique to OLAP operations, *i.e.*, based on the fact that a cube C is visualized, how to efficiently visualize a cube obtained form C through a combination of one or several operations applied to C. We think that this problem is particularly relevant for data exploration, especially when considering the roll-up or drill-down operators, in order to display the most relevant level of granularity of the cube.

References

1. D. Barbara and M. Sullivan. Quasi-cubes: Exploiting approximation in multidimensional data sets. *SIGMOD Record*, 26(3), 1997.
2. L. Bellatreche, A. Giacometti, P. Marcel, H. Mouloudi, and D. Laurent. A personalization framework for olap queries. In *DOLAP '05: Proceedings of the 8th ACM international workshop on Data warehousing and OLAP*, pages 9–18, New York, NY, USA, 2005. ACM Press.
3. L. Cabibbo and R. Torlone. A Logical Approach to Multidimensional Databases. In *Sixth International Conference on Extending Database Technology (EDBT'98)*, pages 183–197, Valencia, Spain, 1998. Lecture Notes in Computer Science 1377,Springer-Verlag.
4. S.K. Card, J.D. Mackinlay, and B. Shneiderman. *Readings in Information Visualization : Using Vision to Think*. Morgan Kaufmann Publishers, 1999.
5. S. Chaudhuri and U. Dayal. An overview of data warehousing and olap technology. *ACM-SIGMOD Records*, 26(1):65–74, 1997.
6. Y.W. Choong, A. Laurent, and D. Laurent. Building fuzzy blocks from data cubes. In *11th IPMU International Conference*, 2006, to appear.
7. Y.W. Choong, D. Laurent, and P. Marcel. Computing appropriate representation for multidimenstional data. *DKE Int. Journal*, 45:181–203, 2001.
8. Y.W. Choong, P. Maussion, A. Laurent, and D. Laurent. Summarizing multidimensional databases using fuzzy rules. In *Proc. 10th International Conference on Information Processing and Management of Uncertainty in Knowledge-Based Systems (IPMU'04)*, pages 99–106, Perugia. Italy, 2004.
9. E.F. Codd, S.B. Codd, and C.T. Salley. Providing olap (on-line analytical processing) to user-analysts: An it mandate. In *White Paper*, 1993.
10. J. Gray, A. Bosworth, A. Layman, and H. Pirahesh. Data Cube: A Relational Aggregation Operator Generalizing Group-By, Cross-Tabs, and Sub-Totals. *Journal of Data Mining and Knowledge Discovery*, 1(1):29–53, 1997.
11. M. Gyssens and L.V.S. Lakshmanan. A Foundation for Multidimensional Databases. In *Proc. 23rd Int. Conf. on Very Large Data Bases*, pages 106–115, Athens, Greece, August 1997.
12. J. Han and M. Kamber. *Data Mining - Concepts and Techniques*. Morgan Kaufmann, 2001.
13. C.G. Healey. *Effective Visualization of Large Multidimensional Datasets*. PhD thesis, University of British Columbia, 1996.
14. W.H. Inmon. *Building the Datawarehouse*. John Wiley & Sons, 1996.
15. M. Jarke, M. Lenzerini, Y. Vassiliou, and P. Vassiliadis. *Fundamentals of Data Warehouses*. Springer-Verlag, 1998.
16. Yahiko Kambayashi, Mukesh K. Mohania, and Wolfram Wöß, editors. *Data Warehousing and Knowledge Discovery, 5th International Conference, DaWaK 2003, Prague, Czech Republic, September 3-5,2003, Proceedings*, volume 2737 of *Lecture Notes in Computer Science*. Springer, 2003.
17. D.A. Keim. Designing pixel-oriented visualization techniques: Theory and applications. *IEEE Transactions on Visualization and Computer Graphics*, 6(1), 2000.

18. D.A. Keim and H.-P. Kriegel. Issues in visualizing large databases. In *Proc. Int. Conf. on Visual Database Systems*. Chapman & Hall Ltd., 1995.

19. R. Kimball. *The Datawarehouse Toolkit*. John Wiley & Sons, 1996.

20. A.S. Maniatis, P. Vassiliadis, S. Skiadopoulos, and Y. Vassiliou. Advanced visualization for olap. In *Proceedings of ACM 6th International Workshop on Data Warehousing and OLAP (DOLAP)*, New Orleans, 2003.

21. G.M. Marakas. *Modern Data Warehousing, Mining and Visualization: Core Concepts*. Prentice Hall, 2003.

22. P. Marcel. Modeling and querying multidimensional databases: An overview. *Networking and Information Systems Journal*, 2(5-6):515–548, 1999.

23. B. Shneiderman. The eyes have it: A task by data type taxonomy for information visualizations. In *Proceedings of the IEEE Symposium on Visual Languages*, pages 336–343. IEEE Computer Society Press, 1996.

24. A. Shoshani. Statistical Databases: Characteristics, Problems, and some Solutions. In *Eigth International Conference on Very Large Data Bases, September 8-10, 1982, Mexico City, Mexico, Proceedings*, pages 208–222. Morgan Kaufmann, 1982.

25. R. Spence. *Information Visualization*. Addison-Wesley (ACM Press), 2000.

26. C. Stolte, D. Tang, and P. Hanrahan. Query analysis and visualization of hierarchically structured data using polaris. In *Proc. of the 8th ACM SIGKDD Intl. Conference of Knowledge Dicovery and Data Mining*, 2002.

27. D. Travis. *Effective Color Displays: Theory and Practice*. Academic Press, London, 1991.

28. P. Vassiliadis. Modeling Databases, Cubes and Cube Operations. In M. Rafanelli and M. Jarke, editors, *Proceedings of the 10th International Conference on Scientific and Statistical Database Management (SSDBM)*, Capri, Italy, July 1998. IEEE Computer Society.

29. P. Vassiliadis and T. Sellis. A survey of logical Models for OLAP Databases. *SIGMOD Record*, 28(4), 1999.

Leveraging Layout with Dimensional Stacking and Pixelization to Facilitate Feature Discovery and Directed Queries

John T. Langton[1], Astrid A. Prinz[2], David K. Wittenberg[3], and Timothy J. Hickey[3]

[1] Charles River Analytics Inc., www.cra.com
jlangton@cra.com
[2] Department of Biology, Emory University
astrid.prinz@emory.edu
[3] Computer Science Department, Brandeis University
{tim, dkw}@cs.brandeis.edu

Abstract. Pixelization is the simple yet powerful technique of mapping each element of some data set to a pixel in a 2D image. There are 2 primary characteristics of pixels that can be leveraged to impart information: 1. their color and color-related attributes (hue, saturation, etc.) and 2. their arrangement in the image. We have found that applying a dimensional stacking layout to pixelization uniquely facilitates feature discovery, informs and directs user queries, supports interactive data mining, and provides a means for exploratory analysis. In this paper we describe our approach and how it is being used to analyze multidimensional, multivariate neuroscience data.

1 Introduction

There are many methods for multidimensional and/or multivariate visualization [1][4][5]. Some popular 2D approaches are scatter plot matrices [1] and parallel coordinates [6]. When faced with the task of visualizing a vast database of neuroscience data we chose to employ a form of pixelization because of the amount of information that can be displayed at one time [7]. This technique maps each data point of some set to a pixel in a 2D image. For our purposes, it meant that we could view an entire dataset of 1,679,616 single compartment neuron simulations in one display. This provided a global context for determining data trends and supporting exploratory analysis.

Keim and Ward have investigated the effects of different layout strategies for multidimensional visualization [17]. In the domain of pixelization, Keim has demonstrated the efficacy of pixel arrangements that are determined by the results of user queries [10] and has proposed algorithms for generating them [8]. In [12], LeBlanc et. al. introduced dimensional stacking which projects multiple dimensions onto two axes so that all are visible in one display.

While LeBlanc et. al. make a subtle reference to the possibility of one data element being mapped to one pixel, the application of dimensional stacking as

P.P. Lévy et al. (Eds.): VIEW 2006, LNCS 4370, pp. 77–91, 2007.

a layout scheme for pixelization has not been thoroughly investigated. We have found that combining both approaches has unique implications for facilitating visual data mining and data trend discovery. In particular, clustering within this scheme is entirely based on the order of dimensions on each axis. The task of interactively permuting dimension orderings can reveal functional dependencies on dimension values and inform user queries for further analysis.

We have applied these methods to the analysis of a highly structured neuron simulation database. The columns that are used as dimensions are independent variables in the simulations and together can be a considered an aggregate primary key. The rest of the columns are dependent variables that are functionally dependent on dimension values. Our visualization approach may not be appropriate for data that has a very different structure. However, these methods have proven quite useful for our purposes and we believe can be generalized further.

In this paper we detail our approach of applying a dimensionally stacked layout to pixelization. Section 2 provides an overview of pixelization, issues of layout, and our generalization and formalization of dimensional stacking in the context of pixelization. Section 3 describes the generation of the data we were tasked to analyze, the motivation for our choice of visualization techniques, and the application of our approaches to that data. We detail the semantics of dimension ordering within the context of our approach in section 4. Both sections 3 and 4 use examples from our experience working with neuroscientists and analyzing multidimensional, multivariate neuroscience data. We specifically show how our methods uniquely facilitate feature discovery, inform and direct user queries, support interactive data mining, and provide a means for exploratory analysis. We conclude our findings in section 5 and provide recommendations for future work in section 6.

2   Background

2.1   Layout

In [17], Ward and Keim describe how layout can convey relations between graphical entities and prescribe placing semantically similar items in close proximity to one another. This poses the issue of how to determine what constitutes semantic similarity. Grouping all red pixels together in a pixelization image may obscure a multimodal distribution of the data values associated with red pixels. It is usually important to retain dimensional context and impart not only attribute values but the dimension values that create them, all in the same visualization.

Keim presents two layout schemes for multidimensional pixelization, both of which map each dimension to a different window [10]. A user can view the same region in 2 or more windows to decipher correlations and functional dependencies between dimensions. In what is termed a query-independent approach, the value of some attribute is mapped to pixel colors, the data is sorted according to that attribute, and then space filling curves are used to arrange the pixels in an image. In a query-dependent approach, an extra window is created with a layout that is driven by a user query. Keim develops a sophisticated metric for determining the distance of a data element from a user query. For the extra query-dependent window, the pixels for items that match the query are placed in the center of the display while the rest are distributed in a generalized spiral pattern, away from the center, in the order of increasing semantic distances. The technique for doing so is described in [8].

The data we were tasked with visualizing contained complicated functional dependencies involving more than two dimensions. To facilitate the visual inspection of these relations we wanted to display both attribute values and all dimension values that created them in the same image. We were therefore unable to use the layout methods described above.

2.2   Dimensional Stacking

This section provides a generalization and formalization of dimensional stacking [12][18], specifically in the context of pixelization. Section 3.2 details our application of this formalization to a set of neuroscience data. Although the combination of these methods was motivated by the structure of our data, our approach can be applied to other data sets as specified below.

One can interpret a relational database as a function on tuples $T = t_1, \ldots, t_n$ where each t_i is a column and T is an aggregate key (a primary key consisting of some number of columns, concatenated). For our purposes, each t_i is considered a dimension and should generally be an independent variable with a small number of possible integer values. The cardinality of possible values for each t_i serves as its base, B (e.g. $B = 2$ if there are two possible values). When the bases of each dimension are equal, T can then be viewed as a base B number with n digits.

For example, suppose $n = 4$ and the dimensions $\{a, b, c, d\}$ can each take on the value 0 or 1. We can then interpret $T = \{a, b, c, d\}$ as a four digit binary number. Leaving the dimensions in alphabetical order, the number 0101 would map to $a = 0$, $b = 1$, $c = 0$ and $d = 1$ or the decimal number 5. It is possible to have an odd number of dimensions and/or dimensions with unequal bases e.g. $0 <= t_i < j$, $0 <= t_{i+1} < k$ and $j \neq k$. These are called mixed radix systems and were first described by Cantor [3]. Knuth presents a comprehensive review of these and related systems in [11]. The algorithms presented below can map to and from a decimal number and any combination of dimensions and dimension bases.

To get a pixelization image we map every combination of dimension values to a set of pixel coordinates x, y. We first partition the dimensions into two non-intersecting sets, X and Y. The x coordinate of each pixel is derived from X and y from Y. Each set can contain any number of dimensions but we generally try to partition the dimensions equally by following the rule: $|X| = n/2$ and $|Y| = n/2 + (n \bmod 2)$ where n is the number of dimensions. The order of the dimensions in X and Y is very significant and is discussed briefly in section 2.4 and in the context of visualizing a set of neuroscience data in section 4.

To continue with the previous example, we can partition the dimensions $\{a, b, c, d\}$ into the two sets $X = \{a, b\}$ and $Y = \{c, d\}$. Given this partitioning, Table 1 shows the placement of each possible combination of dimension values where cells would correspond to pixel locations. The dimensions in each set are then concatenated to form a single number for each pixel coordinate (e.g. ab, cd). Table 2 shows the values of Table 1 with the dimensions in each set concatenated to form a pair of two digit binary numbers for pixel coordinates. Table 3 shows the decimal values for the binary values in Table 2.

In related literature [12], dimensions are described as "faster" or "slower" in reference to the rate at which they increase along the x and y axis. Notice in Tables 1 and 2 that the value for dimension b increases twice from 0 to 1 going left to right along each row. Notice also that the value for dimension a increases from 0 to 1 only once going left to right along each row. Dimension b is therefore "faster" than dimension a because its value increases at a faster rate.

Because we focus on pixelization and concatenate dimension values to form pixel coordinates, we extend the traditional terms of least and most significant digits to least and most significant dimensions. Consider Table 2 to be a pixelization image with a dimensional stacking layout where each cell is a pixel and the origin is at the bottom left. The dimension that makes up the least significant digit in the x coordinate of each cell is b, therefore b is the least significant dimension on the x axis. a is the most significant dimension on the x axis, c is the most significant dimension on the y axis, and d is the least significant dimension of the y axis.

Table 1. Dimension values, split into two sets for x and y coordinates, arranged as $(a, b), (c, d)$ in each table cell

(0,0),(1,1)	(0,1),(1,1)	(1,0),(1,1)	(1,1),(1,1)
(0,0),(1,0)	(0,1),(1,0)	(1,0),(1,0)	(1,1),(1,0)
(0,0),(0,1)	(0,1),(0,1)	(1,0),(0,1)	(1,1),(0,1)
(0,0),(0,0)	(0,1),(0,0)	(1,0),(0,0)	(1,1),(0,0)

Table 2. Dimensions a and b concatenated to form a two digit binary number for the x coordinate and dimensions c and d concatenated to form a two digit binary number for the y coordinate in every cell

00,11	01,11	10,11	11,11
00,10	01,10	10,10	11,10
00,01	01,01	10,01	11,01
00,00	01,00	10,00	11,00

Decimal values can be calculated for pixel coordinates given any partitioning of any number of dimensions of similar or differing bases. The algorithm for doing so is the same as the general algorithm for deriving the placement of "grid squares" shown in [12]. The algorithm for deriving dimension values from a decimal number is given below.

Table 3. Decimal coordinate values derived from the binary coordinates in Table 2

0,3	1,3	2,3	3,3
0,2	1,2	2,2	3,2
0,1	1,1	2,1	3,1
0,0	1,0	2,0	3,0

Given some decimal number m and an ordered set of dimensions $T = t_1, \ldots, t_n$, start with the right most (or least significant) dimension t_n and iterate left through the set computing:

1. $t_i = m \bmod$ (the base of t_i)
2. $m = m/$ (the base of t_i)
3. goto step 1 for $t_{i-1}, \ldots, t_1$

Below we convert the decimal x coordinate for the fourth column of Table 3 where it is 3, into values for the dimensions a and b which each have a base of 2:

1. $3 \bmod 2 = 1$ $(b = 1)$
2. $3/2 = 1$ (ignore remainder, pass 1 on)
3. $1 \bmod 2 = 1$ $(a = 1)$
4. answer = 11 or $(a = 1), (b = 1)$

The width of a pixelization image resulting from this scheme is the product of the bases in X, and the height is the product of the bases in Y. The above algorithms represent a bijection between the x coordinate of each pixel to values for the dimensions in the set X and the y coordinate of each pixel to values for the dimensions in the set Y. Our implementations create only pixelization images and support zooming and panning so do not require a square image nor the blank rows or columns required in [12].

2.3   Pixel Attributes

In pixelization, the color of each pixel is generally determined by some value of the data element corresponding to that pixel. Different color attributes such as hue and saturation can be mapped to different values. If a discrete classification of data items exists then one can associate a distinct color to each class. Continuous attribute values can be normalized and mapped into a spectrum of colors or color gradient. Discrete and continuous values can be combined by mapping each to a different color attribute e.g. hue for continuous values and color for discrete values. Tufte presents well-studied strategies for selecting effective color maps in [16].

Keim shows how his distance measure can be mapped to pixel color intensity in addition to layout [9]. Distance, again, is defined as the relevance of a data item to a user query. We chose to allow user queries to directly define the color map. In our approach queries can return either a discrete or continuous value as described above. In the simple case, a query might test a data item's membership in some class. In a more complicated case the user may compose a query that returns the result of some function on a particular data attribute.

2.4   Dimension Ordering

It is usually considered desirable to cluster graphical displays so that a distinct structure is visible [13][17][19]. In [2], Ankerst et. al. propose clustering dimensions with similar attribute values in order to ensure their collocated arrangement in a pixelization. They further prove that this is an NP-complete problem in the context of certain layout schemes such as space filling curves like Keim's recursive pattern [8]. The computational complexity of finding optimal dimension orders for visual clustering depends upon the layout mechanism being used.

Dimension order has a particular significance for dimensional stacking [13] [18]. There are $n!$ orderings for n dimensions yielding $n!$ images for each color specification. If there are an even number of dimensions with the same cardinality, then half of the images produced are merely the result of swapping the x and y coordinates of each pixel. This is affectively flipping a square image on a diagonal axis. It may also be possible to extract some number of identities where dimensions having different cardinalities can produce square images in certain orderings.

With dimensional stacking, the task of finding dimension orders that reveal informative images is difficult not only because the number of possible orders grows factorially in terms of the number of dimensions, but also because it is impossible to determine what types of clusters or visual patterns the user may find useful. We describe the semantic meaning of dimension order when applying our methods to neuroscience data in section 4. Some automatic approaches are discussed as well as general use patterns.

3   Data

3.1   Motivation for Dimensional Stacking and Pixelization

The impetus for combining dimensional stacking and pixelization to visualize data was borne out of a collaborative effort with colleagues in the Neuroscience department at Brandeis University. In a research effort by Prinz et. al., a large database of model neuron data was generated to better understand how individual membrane currents contribute to the electrical activity of a neuron [14]. The database was generated in a brute force style by several model neuron simulators running in parallel.

Each model neuron was characterized by a sequence of 8 conductance parameters: KCa, Na, CaS, H, CaT, Kd, A, and leak. Each of these conductances could take on 6 possible values and no two models shared the same combination of conductance values. Simulations were run for every combination of parameter values and various attributes recorded for each (e.g. whether the neuron activity pattern was silent, spiking, bursting, irregular). This yielded 6^8 model neuron simulations, each of which correspond to one row in the database tables. The primary key or model number for each of these tables is the model neuron conductance parameter values in the form of an 8 digit base 6 number. The

3GB database contained 2KB of data for each of the 1,679,616 model neuron simulations.

An initial statistical analysis of the database revealed a bimodal distribution of spiking neurons dependent on period length. Some questions arose about the relationship between the parameter values of those neurons and their classification as fast/slow spikers, e.g. what are the locations of these two classes of spikers in conductance space (conductance space refers to the 8 dimensional space of all possible conductance parameter values)? How are they distributed throughout conductance space? What can be said about the border region between these distributions, if there is one?

The need for a visualization tool was evident. Such a tool would not only help with the questions at hand but could also facilitate feature discovery and inform future queries. Two characteristics of the neuroscience database immediately eliminated many candidate tools and visualization approaches:

1. It was vast (1,679,616 data elements with 2KB of information for each).
2. It was multi-dimensional.

XGobi was unable to load a complete table from our database before freezing [15]. Parallel coordinates resulted in overlapping lines obstructing any visual structure [6]. Scatter plots complicated the discovery of functional dependencies that involved more than two dimensions [1]. A requirements analysis was performed for a visualization technique that would be adequately expressive. The desire was to display some attribute value for every model neuron along with the dimension values it was dependent on, all in one image. The size of our data suggested using one pixel per model neuron. The database was generated before we started our work, so the decisions about how many independent variables or dimensions there were and how to discretize their possible values were made for us.

3.2   Applying Dimensional Stacking and Pixelization

In the following sections we present our visualization of the neuroscience database using a combination of dimensional stacking and pixelization. We describe some of the features these techniques provide, however, supporting software tools and implementation specifics are out of the scope of this paper.

Pixel coloring. The database of 6^8 model neuron simulations mapped to 6^8 pixels. Pixel colors were specified by user queries written in SQL. Two types of queries were employed:

1. a select statement that returned a subset of the database containing neuron models that satisfied a particular constraint
2. a function that returned a Real value and was evaluated on an attribute of some number of model neurons in the database

In the first type of query, all pixels associated with the model neurons in the result set were assigned one distinct color. In the second type of query, the pixel associated with each model neuron in the result set was assigned a color gradient determined by the normalized result of the user function. Solid regions of color or repeating patterns represented data trends.

Pixel layout. Each model neuron was uniquely defined by its conductance parameter values, an 8-tuple of integers in the range $[0, 5]$. The conductances were KCa, Na, CaS, H, CaT, Kd, A, and leak and served as the dimensions of our data set. The entire space of possible parameter values and model neurons was therefore 8 dimensional. We reduced this to 2 dimensions by partitioning the parameters into 2 sets of 4 elements each, and viewing them as a pair of 4 digit base 6 numbers. These numbers served as the decimal coordinates of the pixel associated with the model neuron bearing those parameter values. More precisely, we chose a set of 4 conductances (independent variables) x_1, x_2, x_3, x_4 and computed a pixel's x coordinate as:

$$x = x_1 * 6^3 + x_2 * 6^2 + x_3 * 6^1 + x_4 * 6^0$$

The y coordinate was calculated in the same fashion using the remaining conductances. The concise nature of this equation is due to the fact that each dimension has a base of 6. 8 dimensions with 6 possible values partitioned into 2 sets of 4 created a 1296 by 1296 square image. Section 2.2 reviews the algorithms for mapping between dimension values and pixel coordinates for any number of dimensions with any combination of bases. Because pixel coordinates map directly to dimension values, software tools were able to interactively report the dimension values for each neuron model/pixel during mouse over events.

Fig. 1 illustrates four classes of neurons. Those with tonically spiking voltage plots are in red. The blue, green, and purple colors indicate neurons with bursting voltage plots where the number of spikes per second is in the range [0,10], (10,18], or (18,1000] respectively. The white pixels are either silent or irregular bursters; we are not considering those classes in this sequence of images.

The axes of Fig. 1 are labeled with the conductances assigned to each. The most significant dimension on the x axis is KCa and the most significant dimension on the y axis is CaT. Remember that each conductance can take on six possible values. If we break Fig. 1 into a 6X6 grid, each grid square would contain a different combination of values for KCa and CaT with the origin 0, 0 being at the lower left corner. The length and height of one of these grid squares would be the length and height of the lines next to the labels for KCa and CaT respectively. Fig. 3 is a clip that is equivalent to the upper right 6X6 grid square of Fig. 1, where the values for KCa and CaT are both 5. We can also break Fig. 3 up into a 6X6 grid where each constituent grid square contains a different combination of values for Na and Kd. The width and height of these smaller grid squares would be the length and height of the lines next to the labels for Na and Kd. Fig. 4 is a clip that is equivalent to the upper right 6X6 grid square of Fig. 3 where the values for Na and Kd are 5. We can continue with this type of grid decomposition until we have a single 6X6 grid of pixels, where each pixel corresponds to a different combination of values for H and leak.

4   Dimension Ordering

Because we had 8 dimensions (conductance values for our data), there were 8! dimension orderings. By placing 4 dimensions on 2 axes we created square images. Half of the images were therefore identities resulting from swapping the x and y coordinates of pixels or flipping the square images on a diagonal axis. This left us with $8!/2 = 20,160$ possible projections for every set of queries and color assignment.

4.1   Semantic Order Effect

The order of dimensions in dimensional stacking has a profound effect on the resulting image. For instance, the data and color assignments in Figs. 1 and 2 are identical; only the dimension orders vary. One is not necessarily better than the other, but each reveals different information about the underlying data. In Fig. 2 there are very few blue, green, or purple pixels in the lower left corner where H and leak values are 0. Because blue, green, and purple pixels are associated with bursting neurons, this shows that neurons usually do not have a bursting activity when H and leak have low to 0 concentrations. This information is not readily apparent in Fig. 1, however, the red band on the left of Fig. 1 where $KCa = 0$ more clearly shows that there is a predominance of spikers (red pixels) when $KCa < 1$. In particular, the distribution of red pixels primarily in the left and bottom edges of Fig. 1 indicates that spiking neurons only occur when KCa or CaT are relatively low (e.g. less than 4) and that most spiking occurs when at least one of these values is less than 2. One might infer that KCa and CaT therefore have an inhibitory affect on neuron spiking activity.

A somewhat complicated feature is visible in the upper right area of Fig. 1. As the values of CaT and KCa both increase (going up and right diagonally in the image) there are fewer and fewer green and purple pixels. Also, within each square where $KCa > 2$ and $CaT > 1$, the green and purple pixels decrease when the values for Kd and Na decrease. This is apparent in Fig. 3 which is a clip of the upper right corner of Fig. 1 where KCa and CaT are both 5. The degree to which the green and purple pixels decrease as values for Kd and Na decrease, increases as the values for Cat and KCa increase. This suggests that there is a linear relation between these four conductances that exacts a slowing down on the spikes per second of bursting neurons.

4.2   General Use Pattern

In general, the most significant dimensions on each axis of a dimensionally stacked pixelization image determine the clustering of an image. One can perform an interactive and exploratory analysis by reordering dimensions to determine which conductances are dominant in determining a particular phenomenon or pattern. For instance, in Fig. 1 there are white grid points throughout the image wherever $H = 0$. We could reposition H as the most significant dimension of one axis to try and cluster this phenomenon and determine the effects of this parameter. This is

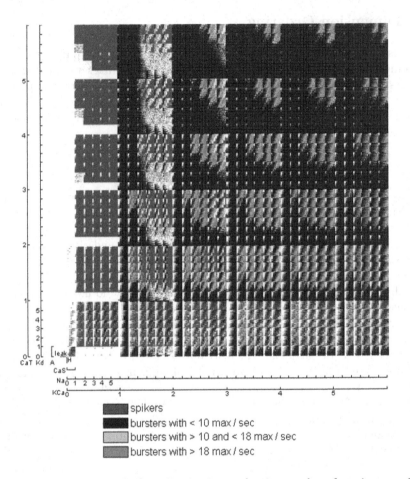

Fig. 1. A dimensionally stacked pixelization image showing number of maxima or spikes per second for bursting neurons, with spiking neurons in red. On the X axis the dimension order is KCa, Na, CaS, and H and on the Y axis the dimension order is CaT, Kd, A, and leak (read from most significant dimension to least significant). Values for less significant dimensions are embedded within each value of more significant dimensions. Only values for the 2 most significant dimensions on each axis are labeled (0-5) for readability.

done in Fig. 2 which shows indeed that there is very little spiking and almost no bursting when H and leak are 0 (the bottom left corner of Fig. 2).

4.3   Automatic Clustering

Within dimensional stacking and pixelization, different dimension orders create different images. While there is no optimal order, images that show clusters or repeating patterns are considered information rich. There is a substantial amount of research on automatic algorithms for finding advantageous dimension

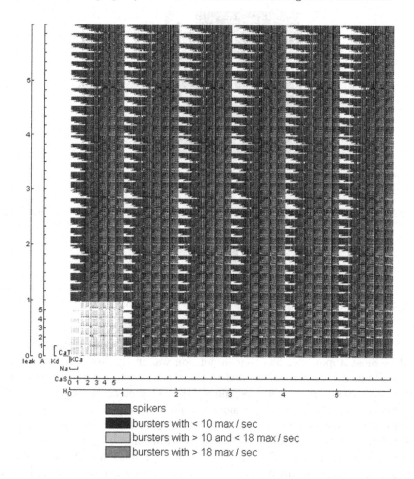

Fig. 2. A dimensionally stacked pixelization image showing number of maxima or spikes per second for bursting neurons, with spiking neurons in red. On the X axis the dimension order is H, CaS, Na, and KCa and on the Y axis the dimension order is leak, A, Kd, and CaT (read from most significant dimension to least significant). Values for less significant dimensions are embedded within each value of more significant dimensions. Only values for the 2 most significant dimensions on each axis are labeled (0-5) for readability.

orders. Yang presents an approach termed DOFSA, an automatic and interactive way of ordering, spacing, and filtering dimensions based on similarity [19]. In their algorithm, dimensions with similar value distributions are clustered and organized into a hierarchical tree structure. Ordering dimensions is then performed as a depth first traversal of the tree. This changes the search problem for finding an optimal dimension order to a sorting problem thereby reducing the computational complexity.

In dimensional stacking, dimension similarity is not necessarily relevant to visual clustering or an advantageous dimension order. In the simple case, a data

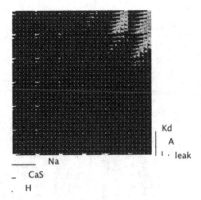

Kd
A
I · leak
Na
_ CaS
. H

Fig. 3. A closeup of the upper right corner of Fig. 1 where KCa and CaT are both 5

A leak
I ·
- CaS
· H

Fig. 4. A closeup of the upper right corner of Fig. 3 where Kd and Na are both 5

trend may be determined by only 1 dimension. In our visualization the trend would be most apparent when that dimension was in the first position on the x or y axis (e.g. its value was assigned to the most significant digit in one of the pixel coordinates). In general, techniques that cluster according to dimension similarity inform us of what dimensions should be grouped together, but do not prescribe a global order. Grouping all pixels of a certain color may also occlude certain data trends such as a multimodal distribution.

Peng defines a measure for the clutter in a dimensional stacked image as the "number of isolated filled bins" / "number of total occupied bins" [13]. For our use of pixelization with dimensional stacking, simply replace "bins" with "pixels." Determining whether a single pixel is connected on any side with another pixel of similar color can be computationally expensive over many millions of iterations. Our neuroscience colleague Dr. Adam Taylor proposed a simpler and more computationally feasible measure of clutter which counts the number of color changes in each row and column of pixels. Using this measure, it is possible to implement algorithms for finding clustered, dimensionally stacked pixelization images.

5   Conclusions

Applying a dimensional stacking layout to pixelization for the analysis of a large neuroscience database provided a number of insights. Analysts could visualize database query results, reorder dimensions to search for data trends, or employ automatic clustering algorithms. As features were discovered, analysts could revise their queries, write new queries, and/or derive parameters for statistical analysis such as linear constraints. Over a period of minutes or hours the conductances of a real neuron may change incrementally. With our visualization, the neuroscientists could trace the activity path of a single neuron through the projection of conductance space in a single image.

We believe our findings have implications for other similarly structured datasets. The combination of dimensional stacking and pixelization provides a visualization that supports the exploratory analysis of multidimensional data that is hierarchical in nature or has a clear delineation of independent and dependent variables. It is also possible to transpose differently structured data into a format conducive for our visualization techniques. The efficacy of such transposition is left for future work. With a dimensional stacking layout, pixelated data, query specified color maps, and interactive dimension ordering, large databases can be interactively explored in an information rich visual environment.

6 Future Work

Much of the background literature on clustering and clutter reduction describes a very small set of heuristic algorithms. One possible area of investigation is to employ methods such as simulated annealing, genetic algorithms, and branch and bound to find information rich dimension orderings. There are many statistical analysis techniques that can inform our visualization, and we believe our visualization can inform the user as to which to use and what to look for. Our methods are not meant to replace traditional data mining techniques but to complement them. A great deal of work can be done in determining best approaches for combining our visualization methods with statistical analysis and data mining algorithms.

Dimensional stacking works best with a small set of dimensions that have monotonically increasing values (and the simplest case is when all dimensions have the same number of values). Our neuroscience data fell within these constraints perfectly, however, without some amount of preprocessing, many other data sets most likely will not. One challenge is to determine what these preprocessing steps might be. Some methodology for determining what data attributes should serve as dimensions and how to discretize their possible values would be necessary. One of our next steps will be to construct such a methodology and further investigate the the general applicability of our approach.

Acknowledgments

We wish to thank Dr. Astrid Prinz for allowing us to use her neuron simulation data, Dr. Eve Marder for introducing us to this problem, and Dr. Adam Taylor for creating a clutter measure and optimization algorithm. Our work on visualization techniques has been propelled forward by many helpful conversations with these neuroscientists.

References

1. D. F. Andrews. Plots of high dimensional data. *Biometrics*, 28:125–136, 1972.
2. M. Ankerst, S. Berchtold, and D. A. Keim. Similarity clustering of dimensions for an enhanced visualization of multidimensional data. In *INFOVIS '98: Proceedings of the 1998 IEEE Symposium on Information Visualization*, page 52, Washington, DC, USA, 1998. IEEE Computer Society.
3. G. Cantor. volume 14. 1869.
4. H. Chernoff. The use of faces to represent points in k-dimensional space graphically. *Journal of the American Statistical Association*, 68:361–368, 1973.
5. S. K. Feiner and C. Beshers. Worlds within worlds: metaphors for exploring n-dimensional virtual worlds. In *UIST '90: Proceedings of the 3rd annual ACM SIGGRAPH symposium on User interface software and technology*, pages 76–83, New York, NY, USA, 1990. ACM Press.
6. A. Inselberg and B. Dimsdale. Parallel coordinates: a tool for visualizing multi-dimensional geometry. In *VIS '90: Proceedings of the 1st conference on Visualization '90*, pages 361–378, Los Alamitos, CA, USA, 1990. IEEE Computer Society Press.
7. D. A. Keim. Designing pixel-oriented visualization techniques: Theory and applications. *IEEE Transactions on Visualization and Computer Graphics*, 6(1):59–78, 2000.
8. D. A. Keim, M. Ankerst, and H.-P. Kriegel. Recursive pattern: A technique for visualizing very large amounts of data. In *VIS '95: Proceedings of the 6th conference on Visualization '95*, page 279, Washington, DC, USA, 1995. IEEE Computer Society.
9. D. A. Keim and H.-P. Kriegel. Visdb: a system for visualizing large databases. In *SIGMOD '95: Proceedings of the 1995 ACM SIGMOD international conference on Management of data*, page 482, New York, NY, USA, 1995. ACM Press.
10. D. A. Keim and H.-P. Kriegel. Visualization techniques for mining large databases: A comparison. *IEEE Transactions on Knowledge and Data Engineering*, 8(6):923–938, 1996.
11. D. E. Knuth. *Seminumerical Algorithms*, volume 2 of *The Art of Computer Programming*. Addison-Wesley, Reading, Massachusetts, second edition, 1981.
12. J. LeBlanc, M. O. Ward, and N. Wittels. Exploring N-dimensional databases. In A. Kaufman, editor, *IEEE Visualization: Proceedings of the 1st conference on Visualization '90*, pages 230–237. IEEE Computer Society Press, 1990.
13. W. Peng, M. O. Ward, and E. A. Rundensteiner. Clutter reduction in multi-dimensional data visualization using dimension reordering. In *INFOVIS '04: Proceedings of the IEEE Symposium on Information Visualization (INFOVIS'04)*, pages 89–96, Washington, DC, USA, 2004. IEEE Computer Society.
14. A. A. Prinz, C. P. Billimoria, and E. Marder. An alternative to hand-tuning conductance-based models: Construction and analysis of data bases of model neurons. *Journal of Neurophysiology*, 90:3998–4015, Dec 2003.
15. D. F. Swayne, D. Cook, and A. Buja. Xgobi: Interactive dynamic data visualization in the x window system. *Journal of Computational and Graphical Statistics*, 7(1):113–130, 1998.
16. E. Tufte. *The Visual Display of Quantitative Information*. Graphics Press, 1983.
17. M. Ward and D. Keim. Screen layout methods for multidimensional visualization, 1997.

18. M. O. Ward. Xmdvtool: integrating multiple methods for visualizing multivariate data. In *VIS '94: Proceedings of the conference on Visualization '94*, pages 326–333, Los Alamitos, CA, USA, 1994. IEEE Computer Society Press.
19. J. Yang, W. Peng, M. O. Ward, and E. A. Rundensteiner. Interactive hierarchical dimension ordering, spacing and filtering for exploration of high dimensional datasets. In *IEEE Symposium on Information Visualization*, page 14, 2003.

Online Data Visualization of Multidimensional Databases Using the Hilbert Space–Filling Curve

Jose Castro and Steven Burns

Costa Rica Institute of Technology
jose.r.castro@gmail.com, royalstream@gmail.com

Abstract. We propose in this paper a visualization approach for large online databases using the Hilbert space–filling curve to map N–dimensional data points to 2D or 3D points. Dimensionality reduction methods like principal component analysis (PCA), multi dimensional scaling (MDS) or self organizing maps (SOMS) can map N–dimensional data points with N>>3 into 3 dimensional or 2 dimensional values that allow us to visualize the data. These methods although popular, require either the calculation of a scatter matrix, eigenvalues and eigenvectors, or the iteration of learning algorithms. Therefore these methods cannot perform online, can be slow with large databases and always produce information loss when the data is mapped from the multidimensional space to the 2D or 3D image. Space–filling curves like the Peano, Z, and Hilbert curve, on the contrary, produce a 1–to–1 mapping between points in a line segment and an arbitrary N–Dimensional hypercube. This 1–to–1 mapping guarantees that there is no information loss on the transformation. Specifically the Hilbert space–filling curve is known to preserve the Lebesgue measure and has been proven to produce an optimal mapping in the sense that an arbitrary contiguous block of information will receive the minimum number of splits in the mapped space. The Hilbert space–filling curve has been extensively used for indexing and clustering by mapping N–dimensional data points to 1–dimensional values. We propose here to use the curve to map to 2 or 3 dimensions for purposes of visualization: By taking advantage of its 1–to–1 nature, a new and generic method to map N–dimensional data points to 2D or 3D points using the Hilbert space–filling curve is developed. We prove theoretically that the calculation of the mapping can be done in constant time if we fix the order of approximation, thereby giving linear O(n) performance on the number of data points to map. We create a Hilbert space–filling curve visualization tool that is much faster than the other methods mentioned and allows us to generate quickly for very large datasets various different visualizations of the data, thereby compensating the lack of use of statistical information in the calculation of the mapped points. We compare our approach to MDS and PCA with a benchmark data set and three real datasets using the distance preserving and topology preserving measure as benchmarks. Our experiments indicate that the Hilbert space–filling curve produces acceptable quality of mapping while achieving much faster visualization and is therefore especially useful for online visualization of very large data sets.

P.P. Lévy et al. (Eds.): VIEW 2006, LNCS 4370, pp. 92–109, 2007.
© Springer-Verlag Berlin Heidelberg 2007

1 Introduction

Dimensionality reduction methods like principal component analysis (PCA) (Duda et al 2000), multi dimensional scaling (MDS) or self organizing maps (SOMS) (Estévez et al 2000) have been extensively used to map N–dimensional data points with N>>3 into 3 dimensional or 2 dimensional values. These techniques have many uses, one of them being the visualization of complex multidimensional data.

Although quite popular, PCA and related approaches require either the calculation of a scatter matrix, eigenvalues and eigenvectors, or the iteration of time consuming learning algorithms, and as a consequence degrade considerably when the number of dimensions and/or data points grows. Another consequence of applying these techniques is that the learning algorithm or PCA matrix compresses the data in a loss-full transformation when the data is mapped from the multidimensional space to the 2D or 3D image (this loss is intentional, since PCA and learning algorithms are meant to extract the meaningful information from the data and eliminate the superfluous features). Also the use of population information for the mapping like the scatter matrix makes it difficult, if not impossible, to create an online visualization of the data points.

In this paper we propose the use of space–filling curves like the Peano, Z, and Hilbert curve for the purpose of data visualization. This approach, like any other, has advantages and disadvantages, and these algorithm trade offs should be known by the visualizer to be able to take good advantage of the technique. We particularly test the Hilbert space–filling curve using the topology preserving measure and the distance preserving measure.

This paper is organized as follows: In the second section we review quickly the principles behind the statistically based techniques and the patter recognition techniques that we will be comparing to (PCA and Sammon Mapping). In the third section we introduce the space filling curves and their different variants focusing specifically on the Hilbert space–filling curve (HSFC). This section ends with a detailed description for the HSFC algorithm that we developed that is a simplified version of Butz original algorithm (Butz 1968). The computation complexity of this method is linear (optimal) with respect to the number of patterns to draw in the data set. Section four explains the distance preserving and topology preserving comparison methods we used and then proceeds in explaining the test sets. Section five presents and discusses the results of our experimentation. We end with conclusions and indications of further research directions that we intend to pursue.

2 Principal Component Analysis

The Karhunen-Loeve transform or Principal Components Analysis (PCA) is a well known linear orthogonal transform widely used in data projection and pattern recognition. In PCA we create a transform that is optimal in a sum squared error sense. PCA first calculates the mean $\overline{\mathbf{x}}$ and scatter matrix S for the data set. Once this is done we calculate the eigenvalues and eigenvectors $\lambda_k, \mathbf{e}_k$ of S. Here, we make a selection of which eigenvectors to use for mapping. Usually the values of the eigenvalues decay rapidly, indicating that some of the dimensions in the multidimensional space comprise noise in the data. For visualization, we select the 2 or 3 largest eigenvalues and their corresponding eigenvectors and create a $d \times k$ matrix $\mathbf{A}$, where d is the number of dimensions and k is the number of chosen eigenvectors (2 or 3). The transformation will be:

$$\mathbf{x}' = \mathbf{A}'(\mathbf{x} - \overline{\mathbf{x}}) \tag{1}$$

2.1 Sammon Mapping and the SAMANN Network

Sammon Mapping (Sammon, 1969) is a useful procedure to reduce the dimensionality of a data set, preserving as well as possible the inter-pattern distances from the original input points.

The distance measure (D) more commonly used is the Euclidean distance and the error function to be minimized is the following:

$$E = \frac{1}{\displaystyle\sum_{i=1}^{n-1}\sum_{j=i+1}^{n} D(\mathbf{x}_i, \mathbf{x}_j)} \sum_{i=1}^{n-1}\sum_{j=i+1}^{n} \frac{\left(D(\mathbf{x}_i, \mathbf{x}_j) - D(\mathbf{y}_i, \mathbf{y}_j)\right)^2}{D(\mathbf{x}_i, \mathbf{x}_j)} \tag{2}$$

That error function is known as the *Sammon Stress* function and Sammon Mapping minimizes it using standard gradient descent or second order methods.

Instead of the standard Sammon Mapping, we will use a Neural Network implementation: SAMANN (Mao, 1995.)

The SAMANN is usually a two layer network with sigmoid activation functions whose input layer has one neuron for each dimension in the input set and the output layer has one neuron for each dimension in the output set. The network is fed two patterns at a time and its trained based on the distance between them with a special learning rule. Please refer to the original paper for more details (Mao, 1995.).

3 Space-Filling Curves and the Hilbert Curve

The discovery of space-filling curves is credited to Peano (1890), when he found a continuous curve that visited every point of a closed square exactly once. A space-filling curve S^n can be considered as a mapping from the unit hypercube $[0,1]^n$ into the unit interval $[0,1]$.

Space–filling curves in general produce a 1–to–1 mapping between points in a line segment and an arbitrary N–Dimensional hypercube (Mokbel et al 2002). This 1–to–1 mapping guarantees that there is no information loss on the transformation.

Specifically the Hilbert space–filling curve is known to preserve the Lebesgue measure and has been proven to produce an optimal mapping in the sense that an arbitrary contiguous block of information will receive the minimum number of splits in the mapped space (Moon et al 2000). The Hilbert space–filling curve has been extensively used for indexing and clustering by mapping N–dimensional data points to a 1–dimensional values (Lawder 2000, Mokbel 2004). We propose here to use the curve to map to 2 or 3 dimensions for purposes of visualization:

By taking advantage of it's 1–to–1 nature, we can map data points from an arbitrary d–dimensional space to a 3 dimensional or 2 dimensional space in two steps: first, for every point $\mathbf{x}$ in the data set, we map it to its one dimensional Hilbert index $\mathbf{x'}$. Once this is done we use this $\mathbf{x'}$ and map it to a value $\mathbf{y} \in [0,1]^3$ using a Hilbert inverse mapping (if we are visualizing in 3 dimensions). This mapping does not lose information because it's 1–to–1 and therefore has an inverse mapping.

In real life, the space-filling curves used are in fact approximations that only visit a finite subset of points by limiting the order of approximation of the curve, but our experience is that with as little as ten bits per dimension we already have a fine enough mapping to be able to give unique image values to every point even for large databases (+2 Million points).

Formally, The mth order approximation curve, denoted by S_m^n, has a grid size of 2^m, and maps a total of 2^{mn} points from an n-dimensional space into a scalar value. The grid size is the number of divisions into which each dimension is split. The actual space filling curve is the limit of this sequence of curves.

$$S^n = \lim_{m \to \infty} S_m^n \qquad (3)$$

Hilbert generalized the definition to an arbitrary number of dimensions and provided a general geometric procedure to construct them. There are many other space-filling curves like the z-curve, the gray curve, etc. However, as shown by (Mokbel et al 2004), the Hilbert space-filling curve produces the least number of splits in an index, as a product of being continuous and devoid of jumps or biased towards any dimension.

The first 3 approximations of the Hilbert curve for a 2-dimensional space can be seen in Figure 1.

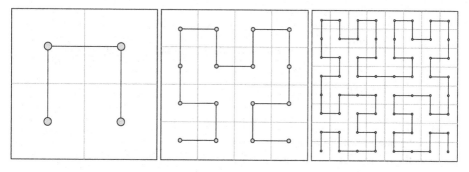

Fig. 1. First 3 approximations for the Hilbert curve in 2 dimensions

3.1 The Hilbert Mapping Algorithm

There are different algorithms for calculating the Hilbert mapping. The one presented is a slight simplification of (Lawder 2000), which in turn is a modification of an iterative algorithm originally presented by (Butz 1968).

The algorithm maps numbers in binary representation and the precision achieved is determined by the order of approximation employed. The output scalar and coordinates of the input vector are real values in [0,1[but the algorithm uses integer variables only with the first bit representing the first bit after the decimal point, so the number 0.11010001 for example would get represented as the integer 11010001.

Please note that all the sub-indexes used in the algorithm are zero-based to ease the programming in any language with zero-based arrays such as C or C++[1].

$$Hilbert(\langle a_0, a_1, ..., a_{n-1} \rangle, m)$$
$$\varpi, \tau, J_{tot} \leftarrow 0$$
$$for\ i \in [0, m-1]$$
$$\qquad \varpi \leftarrow \varpi \otimes \tau$$
$$\qquad \sigma \leftarrow \alpha^i \otimes \varpi$$
$$\qquad \sigma \leftarrow \sigma \lll J_{tot}$$
$$\qquad r^i \leftarrow \sigma \otimes (\sigma \gg 1) \otimes (\sigma \gg 2) \otimes ... \otimes (\sigma \gg (n-1))$$
$$\qquad J \leftarrow principal\ position\ of\ r^i$$
$$\qquad \tau \leftarrow \sigma\ complemented\ at\ position\ (n-1)$$
$$\qquad if\ \tau\ has\ odd\ parity$$
$$\qquad\qquad \tau \leftarrow \tau\ complemented\ at\ position\ J$$
$$\qquad endif$$
$$\qquad \tau \leftarrow \tau \circ\ggg J_{tot}$$
$$\qquad J_{tot} \leftarrow J_{tot} + J$$
$$endfor$$
$$r \leftarrow 0.r^0 r^1 ... r^{m-1}$$
$$return\ r$$

Where:
  n is the number of dimensions.
  m is the order of approximation.

---

[1] Please refer to the Appendix for sample code in C++ for the mapping algorithms.

r is the scalar output of the algorithm (the mapped value) in the range $[0,1]$ represented in $n \cdot m$ bits.

r^i is the ith word of n bits from r with $i \in [0, m-1]$ such that $r = r^0 r^1 ... r^{m-1}$.

a_j is the j_{th} coordinate of the input vector $\langle a_0, a_1, ..., a_{n-1} \rangle$ being mapped represented in m bits, with $j \in [0, n-1]$.

α^i is a word of n bits whose j_{th} bit has the same value as the ith bit of a_j.

$\omega, \sigma, \tau, J, J_{tot}$ are auxiliary variables of the algorithm with no special meaning.

Parity: number of bits of a word whose value is non-zero.

Principal Position: If all the n bits of the word have the same value, the principal position is by definition $n-1$. Otherwise, it's the zero-based index of the rightmost bit whose value differs from the bit in the last position $(n-1)$.

Finally, $\ll$ and $\gg$ are the standard bit-shift operators, $\ll\!\circ$ and $\circ\!\gg$ represent circular bit-shift operations and $\otimes$ stands for the bitwise XOR.

The reverse mapping is the exact same algorithm executed in reverse and it's omitted for the sake of brevity. Nevertheless, the source code for the mapping in both directions is included in the Appendix.

The algorithm is clearly $O(nm)$ which is a clear advantage over the other techniques presented.

4 Comparison Measures

In order to compare numerically the different mapping procedures we use two different measures. The former quantifies the topology preservation of the data whereas the latter measures the conservation of the distances. Both measures provide numerical values for comparison and analysis.

4.1 Topology Preservation

The Topology Preservation Measure (Andreas 2000) provides a way to quantify the conservation of the local neighborhood for each element in the data set.

The index of the ith nearest neighbor for a given pattern $\mathbf{x}_j$ in the original high-dimensional space gets denoted by $NNX(j,i)$. Therefore $\mathbf{x}_{NNX(j,1)}$ would be the nearest neighbor for $\mathbf{x}_j$.

Following the same notation, the ith nearest neighbor for the corresponding pattern $\mathbf{y}_j$ in the low-dimensional space is $NNY(j,i)$. The following credit scheme is then applied:

$$qm_{ji} = \begin{cases} 3 & \text{if } NNX(j,i) = NNY(j,i) & (4) \\ 2 & \text{if } NNX(j,i) = NNY(j,t) & t \in [1,n], \ t \neq i \\ 1 & \text{if } NNX(j,i) = NNY(j,t) & t \in]n,k], \ n < k \\ 0 & \text{else} \end{cases}$$

Basically, each one of the n nearest neighbors of $\mathbf{x}_j$ gets a score between 0 and 3. The highest score means their relative position was preserved exactly by the mapping. The following score applies if their position changed but stayed within the neighborhood of the nearest n elements. Finally, the lowest non-zero score applies if the element is found in a broader neighborhood of k elements. Usual values for n and k are 4 and 10 respectively (Andreas 2000, Estévez et al 2005).

Summing the scores through the whole dataset and dividing by a normalizing factor we obtain the Topology Preservation Measure:

$$qm = \frac{1}{3nN} \sum_{j=1}^{N} \sum_{i=1}^{n} qm_{ji} \qquad (5)$$

As it can be seen from the equation above, the measure is a real number between 0 and 1, where $qm = 1$ would indicate a perfect preservation of the topology for the given parameters.

The topology preservation measure ignores the explicit values of the Euclidean distances and only cares about their relative ordering. Therefore, it's invariant to translations, rotations and uniform rescaling of all coordinates of the data set in one or both dimensional spaces.

4.2 Distance Preservation: Sammon Stress

This measure can be interpreted as an error or penalty assigned to the differences between the distances in the original space and the mapped space, see (2). Note that the first part of the equation is a constant normalizing factor that can be calculated beforehand.

Since this measure is based solely on distances between the points, it's insensitive to translations of the data set and it's said to be invariant to uniform rescaling (D. Ridder 1997).

However, from (2) it can be seen that the measure remains invariant only if both sets of points in both dimensional spaces get rescaled equally and simultaneously. Therefore for a constant set of points in the original high-dimensional space, multiple rescaled versions of given mapping would yield different Sammon Stress measures.

Some dimensionality reduction techniques such as the Hilbert curve or the SAMANN (with sigmoid outputs) produce mapped patterns whose values are restricted to the unitary interval. Rescaling the input data set is not an option because it's not known beforehand how large could get the mapped coordinate values (D. Ridder 1999) and this limits the minimum Sammon Stress they can obtain.

After the mapping process is finished, we propose scaling the output data set by a factor β that minimizes the stress function:

$$\beta = \arg\min_{\beta} \sum_{i=1}^{n-1} \sum_{j=i+1}^{n} \frac{\left(D(\mathbf{x}_i, \mathbf{x}_j) - \beta D(\mathbf{y}_i, \mathbf{y}_j)\right)^2}{D(\mathbf{x}_i, \mathbf{x}_j)} \tag{6}$$

Solving that equation we find the appropriate scaling factor to be applied to avoid unnecessary penalties caused by scale differences:

$$\beta = \frac{\displaystyle\sum_{i=1}^{n-1} \sum_{j=i+1}^{n} D(\mathbf{y}_i, \mathbf{y}_j)}{\displaystyle\sum_{i=1}^{n-1} \sum_{j=i+1}^{n} \frac{D(\mathbf{y}_i, \mathbf{y}_j)^2}{D(\mathbf{x}_i, \mathbf{x}_j)}} \tag{7}$$

The calculation of this factor we propose is $O(n^2)$ which is no different than the calculation of the Sammon Stress itself.

5 Experiments

Three real-world databases of three different sizes were used. For the bigger databases random subsets were taken to run the experiments instead of using the database as a whole because the comparative measures employed, specially the Sammon Stress, are quadratic with respect to the number of cases.

1. *Iris Database:* The Iris database from the UCI Machine Learning Repository is used extensively in pattern recognition and it's also referenced by most of the referenced works about dimensional mapping. It contains information about 150 iris plants of three different classes: Setosa, Versicolor and Virginica. Each element is represented by four different attributes: petal length, petal width, sepal length and sepal width. This database is rather small and we include it solely for comparative purposes.

2. *Letter Image Recognition Data:* Also from the UCI MLR, contains data about the 26 capital letters in the English alphabet. Each letter was rendered in one out of 20 possible fonts and distorted randomly afterwards. The data set contains 20000 different patterns with 16 attributes relative to their position, pixels, dimensions and their means, variances and correlations. Three different data sets of 1000, 2000 and 5000 records were randomly selected to perform the experiments.

3. *Forest Covertype:* From the Department of Forest Sciences of the Colorado State University, contains information about 581012 trees with seven different classes of cover type. The original database presented 54 attributes but number 11 to 14 were mutually exclusive booleans that could get represented as a single integer between

1 and 4. Attributes 15 to 54 presented the same behavior and were also reduced to a value between 1 and 40 leaving us with a total of 12 attributes for this database. Just like before, the experiments were performed over three random data sets of 1000, 2000 and 5000 records.

The high-dimensional data sets were mapped onto a low-dimensional space using the Hilbert Space-filling curve. First each pattern vector is reduced to a single scalar value using the described algorithm. Arbitrary-precision integers were employed in the code preventing any loss of information and allowing this operation to be completely reversible. The scalar value is then expanded using the reverse algorithm to a vector in the 2 or 3-dimensional space for visualization.

This methodology is not employed by any of the reviewed references. It differs from the traditional approach (Keim 1995, Wettenberg 2005) where the Hilbert curve is used to arrange the points of one single dimension into a rectangular area or sub-window, later each dimension gets its own sub-window on the screen and there is no dimensionality reduction taking place whatsoever.

For each one of the data sets, 50 independent runs were performed with a different random order of the dimensions. The Hilbert curve is not biased towards any dimension (Moon et al 2001) therefore similar results are expected from all of them.

For each run, both 2-dimensional and 3-dimensional mappings were created and the Topology Preservation Measure was calculated along with the Sammon Stress after scaling the mapped data according to (7).

Principal Components Analysis was run on the same data sets and the same measures were also calculated for its 2-dimensional and 3-dimensional mappings. PCA is our main point of comparison through the experiments.

Finally, we created a SAMANN network and initialized its first layer with the eigenvector matrix as shown by (Lerner et al 2000). It was run for all the data sets with 1000 or less points and only for a two-dimensional projection. The training was stopped when the change in the error was less than 0.00001 or when a time limit had elapsed.

5.1 Experimental Results

The summarized results for the mappings using the Hilbert curve can be seen in the Table 1. For each of the 50 runs of each data set the best value, the mean and the standard deviation of both quality measures are presented. As expected from the properties of the curve, the standard deviations are low.

The results for the PCA are listed in the Table 2. It gives better topology preservation than the Hilbert curve for smaller databases like the Iris, but underperforms for the Letter-recognition database and the Cover-type database in the two dimensional cases.

Also, for those two bigger databases the quality of the topology preservation measures for PCA seems to decrease as the sample size gets larger. All the statistics of the PCA are based on the $d \times d$ covariance matrix which seems to become less representative of the nature of the data as the number of instances grows.

For the Hilbert mapping the topology preservation doesn't decrease as the size of the data sets increases. Actually for the Letter Recognition database the best topology preservation was obtained when using the biggest sample size of 5000 rows. The obtained values are also surprisingly similar for the 2D and 3D cases.

Table 1. Experimental results using the Hilbert Mapping

Dataset	Sample size	Dims	Topology P.M.			Sammon Stress		
			Best	Mean	StDev	Best	Mean	StDev
Iris	150(all)	2D	0.451	0.430	0.012	0.218	0.338	0.066
Letter-rec	1000	2D	0.222	0.200	0.009	0.233	0.307	0.043
Letter-rec	2000	2D	0.223	0.208	0.006	0.227	0.302	0.042
Letter-rec	5000	2D	0.243	0.234	0.004	0.234	0.305	0.040
Cover-type	1000	2D	0.282	0.271	0.006	0.168	0.392	0.088
Cover-type	2000	2D	0.276	0.266	0.005	0.245	0.386	0.093
Cover-type	5000	2D	0.269	0.262	0.004	0.239	0.382	0.104
Iris	150(all)	3D	0.446	0.416	0.017	0.201	0.281	0.042
Letter-rec	1000	3D	0.224	0.200	0.009	0.185	0.228	0.025
Letter-rec	2000	3D	0.225	0.208	0.007	0.182	0.227	0.030
Letter-rec	5000	3D	0.243	0.234	0.004	0.179	0.223	0.025
Cover-type	1000	3D	0.289	0.272	0.008	0.191	0.311	0.069
Cover-type	2000	3D	0.279	0.267	0.004	0.156	0.300	0.061
Cover-type	5000	3D	0.271	0.264	0.004	0.152	0.293	0.065

Table 2. Experimental results using PCA

Dataset	Sample	Dims	TPM	SS
Iris	150(all)	2D	0.558	0.009
Letter-rec	1000	2D	0.118	0.202
Letter-rec	2000	2D	0.102	0.198
Letter-rec	5000	2D	0.080	0.198
Cover-type	1000	2D	0.267	0.075
Cover-type	2000	2D	0.211	0.074
Cover-type	5000	2D	0.165	0.072
Iris	150(all)	3D	0.782	0.001
Letter-rec	1000	3D	0.262	0.108
Letter-rec	2000	3D	0.243	0.110
Letter-rec	5000	3D	0.205	0.109
Cover-type	1000	3D	0.342	0.045
Cover-type	2000	3D	0.295	0.044
Cover-type	5000	3D	0.262	0.043

PCA shows better performance for the 3D case which was expected because as the number of dimensions in the mapped space grows, the output data set approaches the original data (but on different coordinate axes).

As far as the Sammon Stress, both mapping techniques perform similarly for the Letter recognition database in 2 dimensions, but PCA seems to preserve the distances better for the rest of the cases.

The Hilbert mapping seems to create clusters of points that are very close together and usually share the same class. This could favor class separability but hurts the distance preservation measure.

Figures 2 to 7 show different visualizations obtained by both mapping methods.

As far as the SAMANN network, we found that it gets stuck very easily in local minima and as (D.Ridder 1997) mentions, they are slower and harder to train than ordinary ANNs and there are full papers devoted to its initialization. Nevertheless, for the Iris database it obtained a topology preservation of 0.34 and a Sammon Stress of 0.043, clearly a local-minimum as it can be appreciated in the Figure 8.

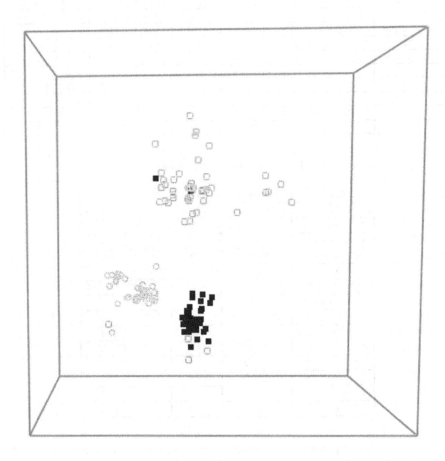

Fig. 2. Hilbert 3D visualization of the Iris database

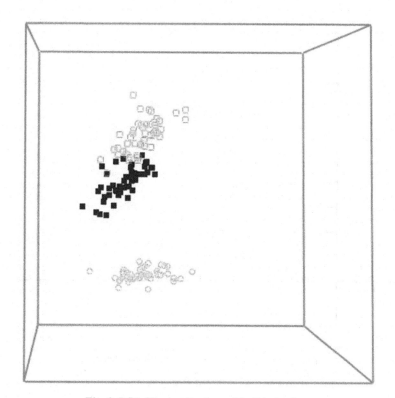

Fig. 3. PCA 3D visualization of the Iris database

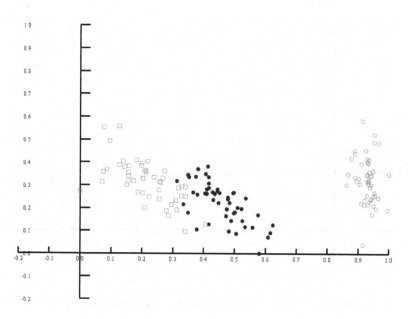

Fig. 4. PCA 2D visualization of the Iris database

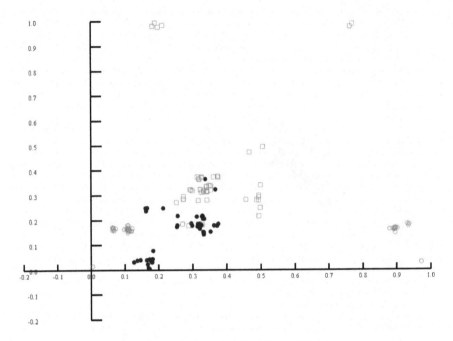

Fig. 5. Hilbert 2D visualization of the Iris database

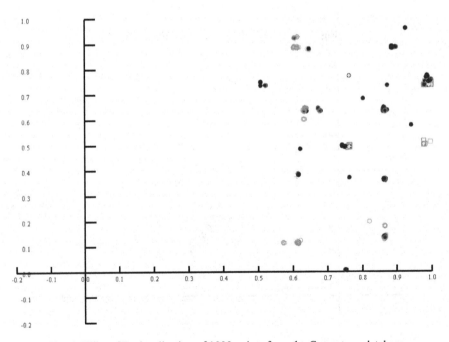

Fig. 6. Hilbert 2D visualization of 1000 points from the Cover-type database

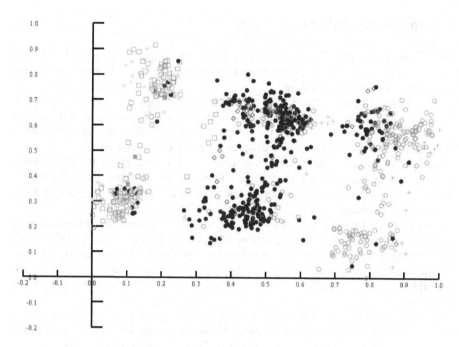

Fig. 7. PCA 2D visualization of 1000 rows from the Cover-type database

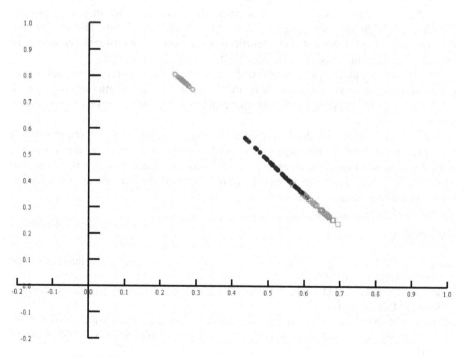

Fig. 8. SAMANN local-minimum visualization for the Iris database

6   Conclusions and Future Work

The Hilbert Curve provides acceptable visualizations at a very small computational cost. The algorithm is linear with respect to both, the number of elements in the data set and the number of dimensions.

Its quantitative performance seems to favor the topology preservation and not so much the distance preservation measure (Sammon Stress). Also, the first measure presented almost no variance for the different runs with different random order of dimensions whereas the Sammon stress presented higher differences.

As an additional contribution, we propose a modification to the distance preservation measure in which the data is rescaled uniformly by a factor that minimizes the error function, allowing us to compare different mapping methods regardless of the output scaling.

As the number of records used in the input data set grows, the Hilbert mapping seems to deal better with the bigger data sets than PCA, this seems to indicate it's better suited to handle large databases than the traditional methods. Large databases also benefit in terms of speed since the mapping is done with integer shift and bitwise operations and the algorithm is linear to both the number of dimensions and the size of the data set.

It's also suited for online databases in which new cases or patterns appear and need to be visualized quickly because the method doesn't involve any recalculation of any statistics and the visualization of each point is independent of the other points in the set.

As a weakness, the mapping using the Hilbert curve treats all the dimensions in the input data set equally which could degrade the quality of the mapping and the visualization if the data set contains noisy or irrelevant dimensions. However, given the simplicity and speed of the algorithm, users could experiment including or excluding certain dimensions as the visualization process takes place.

Visually, most of the projections made with the Hilbert curve show clusters of points of the same class grouped closely. PCA, on the other hand, attempts to preserve the variance of the projected data and this explains the better scores on the Sammon Stress.

Future work might include the use of additional mapping quality measures which take into consideration the separation of the different classes in the projected mapping. Furthermore, this paper is part of a wider ongoing research regarding the applications of the space-filling curves not only to visualization but to classification problems and clustering.

References

(Butz 1968) A. R. Butz. *Space filling curves and mathematical programming*. Information and Control, 12:314-330, 1968.

(Duda et al 2000) Richard O. Duda, P. E. Hart, David G. Stork. *Pattern Classification*. Wiley & Sons. 2nd Edition. 2000.

(Estévez et al 2005) Pablo A. Estévez,; Cristián J Figueroa; Kazimo Saito. *Cross-Entropy Approach to Data Visualization Based on the Neural Gas Network*. IJCNN 2005, Montreal Canada.

(Keim 1995) Keim, D. *Enhancing the visual clustering of Query-Dependent Database Visualization Techniques Using Screen-Filling Curves,* Lecture Notes In Computer Science, Vol. 1183, Springer-Verlag, London.

(König 1998) Andreas König. *A Survey of Methods for Multivariate Data Projection, Visualisation and Interactive Analysis.* Dresden University of Technology, Germany, 1998.

(König 2000) Andreas König. *Interactive visualization and Analysis of Hierarchical Neural Projections for Data Mining.* IEEE Transactions on Neural Networks, Vol. 11, No. 3, May 2000.

(Lawder 2000) J. K. Lawder. *Calculations of Mappings Between One and n–dimensional Values Using the Hilbert Space–filling Curve.* Technical Report no. JL1/00, August 15, 2000.

(Lerner et al (A) 1999) B. Lerner, H. Guterman, M. Aladjem, and I. Dinstein. *A comparative study of neural network based feature extraction paradigms.* Pattern Recognition Letters, vol. 20, no. 1, pp. 7-14, 1999.

(Lerner et al 1996) B. Lerner and H. Guterman and M. Aladjem and I. Dinstein. *Feature Extraction by Neural Network Nonlinear Mapping for Pattern Classification.* ICPR13, Vienna, vol. 4, 320-324, 1996.

(Lerner et al (B) 1999) B. Lerner, H. Guterman, M. Aladjem, I. Dinstein, and Y. Romem. *On pattern classification with Sammon's nonlinear mapping - an experimental study.* Pattern Recognition, vol 31, pp. 371-381, 1998.

(Lerner et al 2000) B. Lerner and Hugo Guterman and Mayer Aladjem and Its'hak Dinstein. *On the Initialisation of Sammon's Nonlinear Mapping.* Pattern Analysis and Applications, Springer-Verlag, London, 2000.

(Mao 1995). Mao, A.K. Jain, *Artificial neural networks for feature extraction and multivariate data projection.* IEEE Trans. Neural Networks 6, 1995.

(Mokbel et al 2002) Mohamed F. Mokbel, Walid G. Aref, Ibrahim Kamel. *Performance of Multi Dimensional Space–Filling Curves.* Proceedings of the 10th ACM symposium on Advances in geographic information systems. ACM Press, New York, USA. 2002.

(Mokbel et al 2004) Mohamed F. Mokbel, Walid G. Aref. Ibrahim Kamel. *Fast and effective characterization of 3D Regions of Interest in medical image data.* Medical Imaging 2004: Image Processing, Proceedings of SPIE Vol 5370. 2004.

(Moon et al 2001) Bongki Moon, H. V. Jagadish, Christos Faloustos, Joel J. Saltz. *Analysis of the Clustering Properties of the Hilbert space–filling Curve.* IEEE Transactions on Knowledge and Data Engineering, Vol 13. No 1. January/February 2001.

(Pekalska et al 1999) Elzbieta Pekalska, Dick de Ridder, Robert P.W. Duin, Martin A. Kraaijveld. *A new method of generalizing Sammon mapping with application to algorithm speed-up.* Delft University of Technology, The Netherlands, 1999.

(de Ridder et al 1997) Dick de Ridder and Robert P. W. Duin. *Sammon's mapping using neural networks: A comparison.* Pattern Recognition Letters, 18:1307-1316, 1997.

(Wattenberg 2005) Martin Wattenberg. *A Note on Space-Filling Visualizations and Space-Filling Curves.* INFOVIS.

Appendix: Sample Source Code

The following C++ source code fragment includes the functions required to perform the dimensional mappings from N-dimensional to 1-dimensional values and back. Please read the remarks section at the end of the source code for additional information.

```
// N-dimensional value to 1-dimensional value:
UINT n_to_one(const UINT a[], int dims, int order)
{
    UINT Wi = 0;
    UINT Ti = 0;
    UINT r = 0;
    int Jtot = 0;
    for(int i = 0; i < order; i++)
    {
        UINT Ai = 0;
        for(int j = 0; j < dims; j++)
            if(is_bit_on(a[j],i))
                Ai = set_bit_on(Ai, j);
        Wi = Wi ^ Ti;
        UINT Oi = Ai ^ Wi;
        Oi = left_shift_circular(Oi, Jtot, dims);
        UINT Yi = Oi;
        for(int j = 0; j < (dims-1); j++)Yi ^= (Oi >> (j+1));
        r |= clean_up(Yi,dims) >> (i * dims);
        int J = get_principal(Yi,dims);
        Ti = swap_bit(Oi,dims-1);
        if(has_odd_parity(Ti,dims))Ti = swap_bit(Ti,J);
        Ti = right_shift_circular(Ti, Jtot, dims);
        Jtot += J;
    }
    return r;
}
  // 1-dimensional value back to N-dimensional space:
void one_to_n(UINT r, int dims, int order, UINT a[])
{
    for(int j = 0; j < dims; j++)a[j] = 0;
    UINT Ti = 0;
    UINT Wi = 0;
    int Jtot = 0;
```

```
for(int i = 0; i < order; i++)
{
    Wi = Wi ^ Ti;
    UINT Yi = r << (i * dims);
    UINT Oi = Yi ^ (Yi >> 1);
    int J = get_principal(Yi,dims);
    Ti = swap_bit(Oi,dims-1);
    if(has_odd_parity(Ti,dims))Ti = swap_bit(Ti,J);
    Oi = right_shift_circular(Oi, Jtot, dims);
    Ti = right_shift_circular(Ti, Jtot, dims);
    UINT Ai = Wi ^ Oi;
    for(int j = 0; j < dims; j++)
        if(is_bit_on(Ai,j))
            a[j] = set_bit_on(a[j], i);
    Jtot += J;
}
}
```

Remarks

- The UINT type found in the code could be a typedef to the platform's 64-bit unsigned integer or simply to unsigned int, but in any event the size of the type imposes a limitation to the number of dimensions (and bits per dimension). For the general case an arbitrary-precision C++ class is required, overloading the shift/bitwise operators and functions accordingly. All our experiments were executed this way.
- This sample code requires the implementation of several bitwise functions such as: is_bit_on, set_bit_on, swap_bit, has_odd_parity, get_principal, clean_up, left_shift_circular and right_shift_circular.
- The last five functions listed require the number of dimensions as a parameter because they need to know how many bits are being used within each UINT value.
- The clean_up function is required to set the remaining (unused) bits to zero when they could interfere with the calculations.

Pixel-Based Visualization and Density-Based Tabular Model

Rodolphe Priam[1], Mohamed Nadif[1], and François-Xavier Jollois[2]

[1] LITA, Université Paul Verlaine-Metz, Ile du Saulcy, 57045 Metz
[2] CRIP5, Université Paris Descartes, 45 rue des Saints-Pères, 75006 Paris

Abstract. Visualization of the massive datasets needs new methods which are able to quickly and easily reveal their contents. The projection of the data cloud is an interesting paradigm in spite of its difficulty to be explored when data plots are too numerous. So we study a new way to show a bidimensional projection from a multidimensional data cloud: our generative model constructs a tabular view of the projected cloud. We are able to show the high densities areas by their non equidistributed discretization. This approach is an alternative to the self-organizing map when a projection does already exist. The resulting pixel views of a dataset are illustrated by projecting a data sample of real images: it becomes possible to observe how are laid out the class labels or the frequencies of a group of modalities without being lost because of a zoom enlarging change for instance. The conclusion gives perspectives to this original promising point of view to get a readable projection for a statistical data analysis of large data samples.

1 Introduction

Increasing number of nonlinear methods $\mathcal{M}$ as [1,2,3,4] are proposed in the literature to project a data sample $\{x_i\}_{i=1}^{i=I}$, demonstrating that the data cloud projection $\{y_i = \mathcal{M}(x_i)\}_{i=1}^{i=I}$ paradigm is a good way to see the structure of the distribution of a dataset. An alternative to this point of view is the self organizing maps (SOM) which construct a tabular projection of the data cloud, with neighbour cells on the table enough similar in the data space to preserve the topology. This way to see data is certainly more readable than a cloud of numerous points on the plane, where points are often hard to see while the shape of the projected distribution appears. For this reason, some zoom tools are generally needed, but can be a way to be lost in the data projection when one has to change the scale of the enlarging. As these new methods are certainly projecting data in a complementary way, and that the tabular view is easily understood, we aim in this paper to add a tabular view for 2D data cloud by two approaches.

A first solution, to get the tabular view, constructs a self organizing map over the bidimensional projection by a constrained Gaussian mixture. To construct a tabular view of a data set, the self-organizing map algorithms are clustering methods with their center vectors constrained with an imaginary regular mesh. Each node of this mesh corresponds to one class center, and the aim of the

P.P. Lévy et al. (Eds.): VIEW 2006, LNCS 4370, pp. 110–118, 2007.
© Springer-Verlag Berlin Heidelberg 2007

algorithm is to train the center vectors such as nearer they are on the mesh, nearer they are in the data space. This strong hypothesis is only true for parts of the mesh because the data distribution is complex in practice. So a SOM algorithm constructs well organized areas separated by a frontier which reveals that these areas are not connected in the data space. To show a tabular view of the projected values y_i, self-organizing maps are an appealing solution. Several algorithms already exist, like the GTM or Generative Topological Mapping [5] for instance. This model is a Gaussian mixture with centers constrained to lay on a discretized surface, and constituting the knot of this discrete surface.

In our case, we are interesting in showing the tabbled data as they are laid out in the plane, as proposed in the tabular model: it is a new constrained Gaussian mixture where means are lying on linear rows and columns which are free to move to show area with high density. Our study permits to construct readable tables from multimedia corpuses. When cells or classes are numerous like in the case of the massive datasets, a Pixel-Based[6] approach is necessary, carrying interesting and synthetic information. In this paper, we present the model, we give the obtained pixel maps for an image dataset, then we explain how visual datamining is beneficial, and finally we conclude with discussion and some perspectives.

2 The Tabular Generative Model

In this section, we suppose that a mapping $\mathcal{M}$ was used to obtain the bidimensional coordinates $y_i = (y_{is})_{s=1,2}$ which are processed to get a tabular view of the projected data cloud. The basis of the algorithm is the Gaussian mixture model which extends the K-means [7] method to a probabilistic expression with a classifying hidden variable. A Gaussian mixture density is written $P(y_i) = \sum_{1 \leq k \leq K} \pi_k G(y_i; m_k, \sigma)$ with K factors, components, or clusters, where the k-th G density is a normal density with the mean m_k and the spherical variance parameter σ. The parameter π_k is the probability that an observation belongs to the k-th component, it therefore corresponds to the proportion of the k-th cluster. The log-likelihood of the observed data $\mathcal{D}$ assumed to be an i.i.d sample $\{y_i\}_{i=1}^{i=I}$ from the probability distribution with density $P(y_i)$ is given by:

$$\mathcal{L}(\theta|\mathcal{D}) = \sum_{1 \leq i \leq I} \log \left\{ \sum_{1 \leq k \leq K} \pi_k G(y_i; m_k, \sigma) \right\},$$

where $\theta = (m_1, m_2, \cdots, m_K, \pi_1, \pi_2, \cdots, \pi_{K-1}, \sigma)$. Inference of this model is done by maximizing the loglikelihood which is intractable in an exact closed form solution. The EM algorithm[8] is applied to this problem by using the likelihood completed by the knowledge of the partition $\mathcal{Z} = \{\mathcal{Z}_1, \mathcal{Z}_2, \cdots, \mathcal{Z}_K\}$,

$$\mathcal{L}(\theta, Z|\mathcal{D}) = \prod_{1 \leq i \leq I} \pi_{z_i} G(y_i; m_{z_i}, \sigma)$$

where z_i the latent variable of which the unknown value is in $\mathcal{K} = \{1, 2 \cdots, K\}$.

It proceeds iteratively in two steps, E (for expectation) and M (for maximization), in maximizing the conditional expectation of $\mathcal{L}(\theta, Z|\mathcal{D})$, given a previous current estimate $\theta^{(t)}$:

$$Q(\theta|\theta^{(t)}) = \sum_{1 \le i \le I} \sum_{1 \le k \le K} q_k^{(t)}(y_i) \log \left\{ \pi_k G(y_i; m_k, \sigma) \right\}$$

$$= \sum_{1 \le i \le I} \sum_{1 \le k \le K} q_k^{(t)}(y_i) \left\{ \log \pi_k - \frac{||y_i - m_k||^2}{2\sigma^2} \right\} - I \log \sigma + cst$$

where

$$q_k^{(t)}(y_i) = \frac{\pi_k^{(t)} G(y_i; m_k^{(t)}, \sigma^{(t)})}{\sum_{1 \le \ell \le K} \pi_\ell^{(t)} G(y_i; m_\ell^{(t)}, \sigma^{(t)})}$$

denotes the conditional probability, given $\mathbf{y}$ and $\theta^{(t)}$, that y_i arises from the mixture component with density $G(y_i; m_k^{(t)}, \sigma^{(t)})$. Each iteration of EM uses the following steps:

- E-step: compute the conditional expectation of $\mathcal{L}(\theta, Z|\mathcal{D})$; in the mixture case this step reduces to the computation of the conditional density $q_k^{(t)}(y_i)$.

- M-step: compute $\theta^{(t+1)}$ maximizing $Q(\theta, \theta^{(t)})$ which leads to:

$$\pi_k^{(t+1)} = \frac{1}{I} \sum_{1 \le i \le I} q_k^{(t)}(y_i),$$

$$m_k^{(t+1)} = \frac{\sum_{1 \le i \le I} q_k^{(t)}(y_i) y_i}{\sum_{1 \le i \le I} q_k^{(t)}(y_i)},$$

and $\sigma^{(t+1)} = \sqrt{\frac{1}{I} \sum_{1 \le i \le I} \sum_{1 \le k \le K} q_k^{(t)}(y_i)||y_i - m_k^{(t+1)}||^2}.$

In the following subsection, we constraint the m_k vectors to get a matrix representation.

2.1   The Model

We construct a regular mesh with K_1 columns and K_2 rows, such as k is indiced by $k = K_1 \times (k_2 - 1) + k_1$, $(k_1 = 1, \ldots, K_1; k_2 = 1, \ldots, K_2)$, and $K = K_1 K_2$. Let's have:

$$m_k = \begin{bmatrix} m_{k,1} \\ m_{k,2} \end{bmatrix} \text{ with } m_{k,s} = \frac{\sum_{1 \le \ell \le k} exp(u_{\ell,s})}{\sum_{1 \le \ell \le K_s} exp(u_{\ell,s}) + 1},$$

where $s \in \{1, 2\}$, and $u_{k,s}$ is a real unknown parameter which can be obtained in maximizing $Q(\theta|\theta^{(t)})$. Our parametrization is such as $0 \le m_{k,s} \le 1$ to get simpler expressions. Note that, the additive constant 1 in $m_{k,s}$ can be replaced by any positive real, and the sum over the numerator induces a fixed topological direction for the corresponding probabilities: a topology-driven variant of the classical *soft-max* [9] parameterization. Finally, we normalize the range of the vector components by calculating $\tilde{y}_{is} = (y_{is} - min_i(y_{is}))/(max_i(y_{is}) - min_i(y_{is}))$. In this case, with $q_k^{(t)}(y_i) = q_{(k_1,k_2)}^{(t)}(y_i)$ and $\pi_k = 1/K$ -roughly speaking, to put more centers in high density areas-, we propose the new marginal-sensitive criterion:

$$\mathcal{Q}_m(\theta|\theta^{(t)}) \equiv \sum_{1\leq s\leq 2} \sum_{1\leq i\leq I} \sum_{1\leq k\leq K_s} q_{k_s}^{(t)}(y_i)(\tilde{y}_{is} - m_{k,s})^2,$$

where $q_{k_s}^{(t)}(y_i) = q_{k,s}^{(t)}(y_i)$ is the marginal of $q_{(k_1,k_2)}^{(t)}(y_i)$ over k_1 if $s = 2$, and k_2 if $s = 1$. A closed form for maximizing this quantity does not exist yet, so we use a gradient ascent to calculate:

$$m^{(t+1)} = \text{argmax}_m \mathcal{Q}_m(\theta|\theta^{(t)}).$$

By derivating the criterion, we get the gradient vector $\mathbf{Dm}^{(t)}$ with the $u_{\ell,s}$ derivative components:

$$\mathbf{Dm}_{\ell,s}^{(t)} = \sum_{1\leq i\leq I} \sum_{1\leq k\leq K_s} q_{k_s}^{(t)}(y_i)(m_{k,s}^{(t)} - \tilde{y}_{is})(m_{\ell,s}^{(t)} - m_{\ell-1,s}^{(t)})(\delta_{\ell\leq k} - m_{k,s}^{(t)}),$$

with $m_{0,s} = 0$, and $\delta_{\ell\leq k}$ is one if $\ell \leq k$ and zero else. Finally, we end to the gradient step:

$$u_{\ell,s}^{(t+1)} = u_{\ell,s}^{(t)} - \rho^{(t)} \mathbf{Dm}_{\ell,s}^{(t)},$$

for a well chosen $\rho^{(t)}$ decreasing function to get a converging value for our $u_{\ell,s}$ parameters. A Newton-Raphson step by calculating the Hessian is an alternative; its expression is not given here. Finally, iterating calculus of the $u_{\ell,s}$ and σ converges towards a maximum (local). We note with a hat the final parameters. The resulting centers, in the data space, are easily found from the $\hat{m}_k$ vectors by the inverse preceding translation, homothety, and a possible rotation when needed, as explained below.

2.2   The Rotation

When the variance data is not very well oriented compared to the axes of the reference mark, one can prefer to carry out to a rotation to get most of the inertia for the tabular view. A former approach to deal with the choice to optimally rotate the tabular view along most of the data variance directions is to add identical non spherical covariance matrix in the Gaussian densities. This no explicit solution is replaced here by an ad hoc rotation over the data sample. We add a matrix transformation on $\tilde{y}_i$ such as we replace $\tilde{y}_i$ by $\bar{\tilde{y}} + W(\tilde{y}_i - \bar{\tilde{y}})$ where W is a $\mathbb{R}^{2\times 2}$ matrix and $\bar{\tilde{y}}$ is the empirical (sample) mean of the vectors $\tilde{y}_i$. The

gradient is updated by replacing $\tilde{y}_{is}$ by $\bar{\tilde{y}}_s + \sum_{s'} w_{ss'}^{(t)} \bar{\bar{\tilde{y}}}_{is'}$ where $\bar{\bar{\tilde{y}}}_{is'} = (\tilde{y}_{is'} - \bar{\tilde{y}}_{s'})$.
Moreover, each EM iteration induces a new matrix $W^{(t+1)}$. A more elegant and
general transformation is the true rotation matrix $W^{(t)}$, written, for the rotation
angle $\alpha^{(t)}$, and an implicit proportional scaling β for $\bar{\tilde{y}}$:

$$W^{(t)} = \begin{bmatrix} cos\,\alpha^{(t)} & -sin\,\alpha^{(t)} \\ sin\,\alpha^{(t)} & cos\,\alpha^{(t)} \end{bmatrix}.$$

So, we get the 1-st derivative $\mathbf{D}_\alpha^{(t)}$ and 2-nd derivative $\mathbf{H}_\alpha^{(t)}$ formulas which
provide the Newton-Raphson update step:

$$\alpha^{(t+1)} = \alpha^{(t)} - \mathbf{D}_\alpha^{(t)}/\mathbf{H}_\alpha^{(t)},$$

ending to a new matrix $W^{(t+1)}$. The complete enhanced model shares interesting
properties with several approaches from data analysis as explained in the next
illustrating section.

3   Application to a Pixel-View for Real Datasets

The tabular model is similar to a discretized[10] linear principal component[11]
from the projected data cloud which gives almost the same rotation: when the centers are enough numerous, and with an unsmoothing affectation of the data vectors
to classes. The corresponding solution is similar to the PCA projection because:

$$\sum_i ||\tilde{y}_i - [\bar{\tilde{y}} + \hat{W}^{-1}(\hat{m}_{\hat{z}_i} - \bar{\tilde{y}})]||^2 = ||\tilde{y}_i - \hat{W}^{-1}\hat{m}_{\hat{z}_i}||^2$$

is then minimal. The method is connected also to a discretization approach and
a density histogram construction for bivariate continuous data: we obtain two
new discrete variables, along rows and columns, with interval choice sensitive
to the marginal density of the bivariate projection. We are able to see the data
cloud in an easier and quicker readable way. It is a complementary approach to
the existing solutions in visualization, with the strong perspective to ameliorate
their own definitions thanks to the help of the local density of the data cloud.
We illustrate the tabular view on 500 images from the dataset[1] of 2000 images
from handwritten digits digitalized in binary images. The projecting method is
here the Locally Linear Embedding or LLE mapping [3]. This method is very
efficient when dealing with images database which contains classes of objects
with similar shapes because locally linear relations exist. We get the resulting
projection or $\tilde{y}_i$ sample in Figure 1. The tabular mapping is able to show the
projective discretized plane as a set of pies in Figure 5. Each cell corresponds
to one cluster with center $\hat{m}_k$, and the pie inside gives the relative frequencies
among the 10 classes of digit (0-9).

It is clear that such a view is very interesting for moderated size table. One is
able to explore the data cloud on the plane without taking care of the exact relative plot positions or local density change which should have induced a possible
annoying variation of the mental map. When the number of rows and columns

---

[1] ftp://ftp.ics.uci.edu/pub/machine-learning-databases/mfeat/

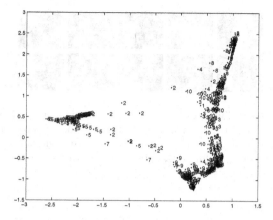

Fig. 1. The LLE projection of the 500 digits with 6 neighbours

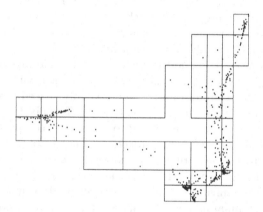

Fig. 2. The 500 digits projection with the rectangular clusters from the Tabular Model

grows up, we need a new way to see how classes are organized among the cells because the table is too large. A pixel view is then very natural and brings us to a synthetic and possibly evolutive or interactive visualization. Giving the location of the ten classes over the discretized plane in Figure 2, each class separately by a map of bits, we get the well-defined visualization in Figure 3.

We are able to compare the location of the classes over the maps, without the need to memorize some preceding map or superpose complex continuous bivariate densities as it would be the case with the classical solution of the smoothed visualisation of the original projected data cloud. Some additional information can even be put over the pixel-base level. For instance, considering one unique map we can provide several colours for the class inside only one cell. It is also possible to replace the label classes by the values of a group of variables. This solution is very interesting for textmining where contextual meaning exists among words for instance.

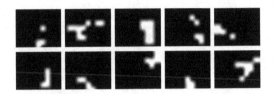

Fig. 3. The 10 bitmaps illustrate the pixel-view paradigm: they show in a synthetic fashion the relations, for LLE, between the 10 classes of digits. A potentially interactive application is to select an area from the original 2D projection and then discover from the pixel maps the corresponding classes intensities with colour.

4   Conclusion

We have proposed a new way to explore a data projection by a new generative model which constructs a table from a cloud which is generally difficult to read when data are too numerous. Our method has several outstanding properties which permit it to carry out several interesting operations over the projection and to get a final interpretable representation.

Roughly speaking, our method is similar to a PCA rotation preceding by a bivariate discretization near a percentile-based one which can be used to initialize the parameters. This is an alternative to an equally-spaced cut of the plane to provide us most of the visual information to show. The pixel-based visualization is introduced to show the final maps and to allow their comparison all together. The method is independent of the projection choice to obtain the 2D plots. It is an alternative to the self-organizing map over the original dataset, when a bidimensional projection yet exists by any nonlinear method.

Moreover, a self-organizing map could have been used to analyse the 2D projection instead of the tabular model. The difference is the risk to loose the exact plot locations over the original projection. This alternative approach is complementary to our work and demands to be studied in the future. Some additive constraints can be used to enhance the tabular contents. It seems that the use of the density of a projected data cloud is under-used in the current tools for visualizing massive datasets. We believe that such information can be further used instead of the classical 3D continuous bivariate density in particular. Several extensions of our approach are possible too, like new ways to see the local density of the data cloud, new ways to zoom a part of the plane, and to take advantage of the pixel paradigm.

References

1. J. Sammon, A nonlinear mapping for data structure analysis, IEEE Transactions on Computers 5 (18C) (1969) 401–409.
2. P. Demartines, J. Herault, Curvilinear component analysis: A self-organizing neural network for nonlinear mapping of data sets, IEEE Transactions on Neural Networks 8 (1) (1997) 148–154.

3. S. T. Roweis, L. K. Saul, Nonlinear dimensionality reduction by locally linear embedding, Science 290 (5500) (2000) 2323–2326.
4. X. He, D. Cai, W. Min, Statistical and computational analysis of locality preserving projection, in: The 22nd International Conference on Machine Learning (ICML2005), 2005, pp. 281–288.
5. C. M. Bishop, M. Svensén, C. K. I. Williams, Gtm : A principles alternative to self-organizing map, Advances in Neuronal Processing System 9, MIT Press.
6. D. Keim, Information visualization and visual data mining, IEEE Transactions on Visualization and Computer Graphics 7 (1).
7. J. MacQueen, Some methods for classification and analysis of multivariate observations, in: 5th Berkeley Symp. Math. Stat. and Proba., Vol. 1, 1967, pp. 281–296.
8. A. Dempster, N. Laird, D. Rubin, Maximum-likelihood from incomplete data via the em algorithm, J. Royal Statist. Soc. Ser. B., 39.
9. C. M. Bishop, Neural Networks for Pattern Recognition, Clarendon Press, 1995.
10. J. C. Fu, L. Wang, A random-discretization based monte carlo sampling method and its application., Methodology and Computing in Applied Probability 4 (2002) 5–25.
11. L. Lebart, A. Morineau, K. Warwick, Multivariate Descriptive Statistical Analysis, J. Wiley, 1984.

Appendix

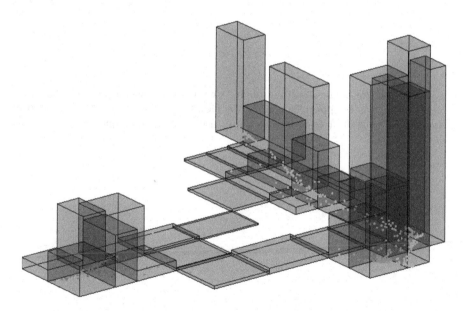

Fig. 4. The 3D histogram according to the Tabular Model: the model clusters the plane in several rectangular areas to help browsing ; each box-voxel has a volume which is proportional to the a posteriori probability of its corresponding center such as the local density of the projection appears

Fig. 5. An example of the resulting tabular view using a table of pies. The local frequencies of the classes appear by reading the pies, so it is easy to see the hidden local relations inside the data according to the projection without requiring to zoom the high density areas.

Pixelization Applications

A Geometrical Approach to Multiresolution Management in the Fusion of Digital Images

Julien Montagner[1], Vincent Barra[2], and Jean-Yves Boire[1]

[1] ERIM, Faculty of Medicine, BP 38, 63001 Clermont-Ferrand Cedex 1, France
{julien.montagner,j-yves.boire}@u-clermont1.fr
[2] LIMOS, Scientific Complex, Les Cézeaux, 63177 Aubière Cedex, France
vincent.barra@isima.fr

Abstract. In most image fusion-based processes, image information are qualified by both numerical activities held by pixels or voxels (data domain) and spatial distribution of these values (spatial domain). Image data are often transformed (registration, multi-scale transform, *etc.*) early in the fusion process, thus losing a part of both their physical meaning and their numerical accuracy. We propose here a new image fusion scheme in which spatial information are managed apart from image activities, aiming at delaying the alteration of original data sets until the final aggregation/decision step of the process. The global idea is to independently model image information from the data and spatial domains, design fusion operators in both domains, and finally obtain the image aggregation model by combining these operators. Such a process makes it possible to introduce spatial coefficients resulting from spatial fusion into advanced aggregation models at the final step. The fusion in the spatial domain is based on discrete geometrical models of the images. It consists in applying a computational geometry algorithm stemming from the study of the classical digital coordinates changing problem, and modified to be efficient even on large 3D images. Two applications of the fusion process are proposed in the field of medical image analysis, for brain image synthesis and activity quantification, mainly destined to the automated diagnosis of Parkinsonian syndromes.

1 Introduction

The principle of data fusion is widely used in image-based processes where the data are acquired from several sources, *e.g.* in the fields of satellite or biomedical imaging [1]. Data stemming from one source are generally used to compensate for a lack of information or as a medium for complementary features about the studied physical object or phenomenon. In this paper, image fusion refers to a computer-based process aiming at extracting knowledge from the image set, which was obviously visible in no original image. These new information may be either qualitative (visual result) [2], quantitative (numerical indexes qualifying the studied subject) [3], or even consist in a subset of image regions (image segmentation in *e.g.* binary sets) [4].

P.P. Lévy et al. (Eds.): VIEW 2006, LNCS 4370, pp. 121–136, 2007.

1.1   Fusion of Discrete Data Sets

Digital images stemming from such acquisition processes are composed by discrete sets of numerical values (2D/3D arrays of image activities), standing for given features of the measured physical phenomenon. A pixel (or voxel) from one of these images is viewed here as the region of the real space in which the associated value has been quantified [5]. In the case where all images have the same size and spatial resolution, the fusion process can sometimes be simply performed by aggregating image activities [6] associated with a unique pixel (with the same index in image arrays). This is only possible if acquisitions have been made from the same point of view on the studied object, and if this one is represented with the same size in each image.

Since this configuration rarely appears (because of practical acquisition constraints, *e.g.* sensor size, spatial and temporal resolutions), most numerical activities in a given image don't have direct spatial correspondence with activities from other data sets (see fig. 1). Both the numerical aspects (data domain) and the spatial distribution of image activities (spatial domain) have therefore to be managed in the data fusion process, each of these features being of equal relevance in the definition of image information. This work typically concerns the fusion of images with different spatial resolutions, usually providing new information at the higher resolution. In the following, the proposed method is explained in the case of two input images (denoted by high/low-resolution images), although it can be applied to a more general fusion problem.

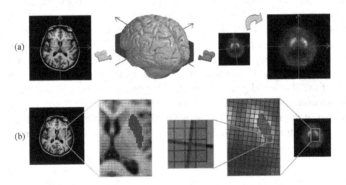

Fig. 1. Image slices of a human brain (a), obtained by both an MR (left) and a SPECT (right) acquisition systems. The SPECT image is resampled to match the other one. Line (b) shows the consequences of these spatial variations on original image activities and pixels (here shown as gray level square areas) through the example of an anatomical structure (putamen).

1.2   Fusion in the Data/Spatial Domains

Whether the images are spatially aligned or not, their activities are often not directly compatible, owing to their numerical nature (numerical type, value range, *etc.*) and, above all, their physical meaning. Aggregating the information held by two corresponding pixels thus implies to model these information in a common formalism.

In this study, the theory of fuzzy sets [7] has been chosen as a mathematical framework for image fusion [1, 8]. A set of numerical degrees is assigned to each pixel, describing to what extent this pixel verifies the studied properties, and thus expressing its membership to the related logical sets (*e.g.* to gray/white matter in a brain MR image). Membership degrees are processed from the image activities and their properties, and are thus a derived form of the image information, versatile enough to allow the management of inaccuracy and uncertainty thanks to fuzzy logic operators [6]. Moreover, this new support forms membership maps for each image/logical set couple, which can be handled together for the information fusion in the data domain.

The question of information modeling and aggregation in the data domain using fuzzy sets or other uncertainty models has been widely studied [4, 6, 9], while the problem of spatial matching of data sets to fuse is often considered of secondary importance.

In order to manage both geometrical relations (only rigid transforms are addressed here) and the difference in spatial resolution between images, most methods process the image information in a multiscale context. Such processes are either based on frequency analysis of data, *e.g.* managing all image information on a common wavelet base [10], or on a "resolution hierarchy" obtained by iterative degradation of original images [11]. The problem can be reduced to the trivial fusion case presented above (direct spatial correspondence of image activities) by simple registration (see fig. 2) of image data in a common geometrical base [12], including an interpolation of numerical activities.

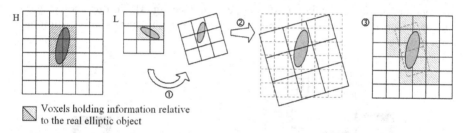

Voxels holding information relative to the real elliptic object

Fig. 2. 2D illustration of the image rigid registration process. The registration transform is composed of a rotation/translation (1) and a rescaling (2) of low-resolution image data (L) to fit the high-resolution image (H). During the last step, the transformed L data are finally interpolated (3).

In each case, image data are transformed at an early stage in the fusion process, and lose a part of both their physical meaning and their accuracy (numerical accuracy when the low-resolution image is resampled, spatial accuracy in the other case). This kind of results is often considered as satisfactory, since most image fusion processes aim at introducing missing spatial or frequential components in a given image for display purpose. However, early alteration of image activities may influence the whole fusion process (successive rounding, *etc.*) and finally lead to critical changes in quantitation result, in the second case of fusion process.

1.3 Proposed Fusion Scheme

The proposed fusion method aims at combining image data really coming from the same spatial location, while preserving original information from each data set until the final aggregation step of the fusion process [1]. In this context, spatial features of the images cannot be managed together with values of the data domain (*e.g.* data interpolation – see fig. 3).

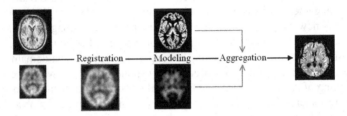

Fig. 3. Usual image fusion process in the context of medical imaging. Fusion of an MR and a perfusion SPECT image of the human brain to create a new image of *e.g.* hypoperfused gray matter. The registration step involves the interpolation of SPECT activities at the beginning of the process.

On the opposite, we propose to independently model image information from the data and spatial domains (avoiding the 3rd step of the registration process presented on fig. 2), design fusion operators in both domains, and finally obtain the image aggregation model by combining these operators (see fig. 4). To our knowledge, no similar fusion method have been proposed yet to manage spatial information apart from image activities.

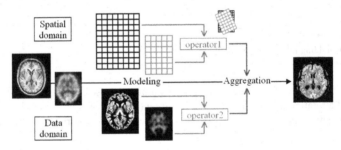

Fig. 4. Proposed fusion scheme. Fusion of an MR and a perfusion SPECT image of the human brain to create a new image of *e.g.* hypoperfused gray matter. Image data and spatial features are managed in parallel during the whole process, and original information are only transformed at the final aggregation step, using a global fusion operator built from operators 1 and 2.

The next section of this paper gives a more formal description of this new image fusion scheme in a general context. Paragraphs 3 and 4 explain how the spatial model can be implemented in the case of 3D images. Part 3 describes the image model itself and explains how it is built using discrete geometry principles. This paragraph also refers to a computational geometry algorithm used to perform the fusion in the spatial domain. Part 4 addresses the question of running performances of a fusion process based on this geometrical model, and explains how to transform the model to improve the processing time.

Finally, we present two examples of image fusion in the context of medical imaging. The fusion scheme is applied (using the 3D geometrical model) to both the synthesis of a diagnostic information from MR and SPECT images of the human brain (qualitative information), and the quantitation of neuronal activity in brain SPECT images using segmented anatomical structures as measurement regions (quantitative information).

2 Methodology

The center of the proposed method is the geometrical model used to represent spatial features of input images, and to perform the fusion in the spatial domain. Indeed, the part of the process achieved in the data domain leads to information models and fusion operators such as those presented in section 1.2. Central questions raised when designing the spatial model are:

- Which kind of "spatial" information are required by the final aggregation step, when merging operators from both the spatial and the data domains? Which form is it possible to give to these information in order to blend the resulting values with those stemming from the data domain?
- Which spatial features of input images are required to process such information, and how to represent them in the spatial model?
- Which kind of fusion operator (operator1 – see fig. 4) can be applied to spatial models to obtain information defined in the first point?

2.1 Form of the Spatial Information

The final fusion we propose to perform is called "activity redistribution". In the spatial domain, it consists in modeling spatial features and relations between both images by simple sets of numerical values (spatial coefficients). These values are thereafter used as weighting factors in the assignment of data from one image or data model to another (usually from low-resolution to high-resolution). The major difference between the activity redistribution and a simple interpolation of image data is the ability to introduce the spatial coefficients into advanced aggregation models at the final aggregation step.

2.2 Choice of the Geometrical Model

Spatial coefficients have to express spatial relations between image data, in order to combine data really coming from the same spatial location. Image activities are obtained from integration of the measured phenomenon in regions of the real space corresponding to image voxels [5] (both the proposed geometrical model and described examples are 3D, but the general method also applies to the case of 2D images and pixels). To study the spatial distribution of image activities and spatial relations linking one image with the others, the proposed model thus simply consists in the representation of influence regions of each numerical activity during the measurement, *i.e.* models of the image voxels themselves. Modeling the whole image thus boils down to define a tilling of the image space by these voxel models.

2.3 Principle of the Activity Redistribution

Let $\rho(v,V)$ be the spatial coefficient processed from the fusion between spatial information linked with voxels v and V. v refers to a "small-size" voxel from the high resolution image H, and V to a "high-size" voxel from the low resolution image L. The elementary space region v is also involved in the structure/composition of the fusion image/result, since this one is designed to have the same spatial resolution as H. The contribution φ_v of voxel v to the final fusion result Φ is possibly processed as:

$$\varphi_v = f(v) + \delta_v \tag{1}$$

where f is a function of v (*e.g.* linked with the numerical activity H(v) of v in H), and δ_v is the redistributed form of the low-resolution information from L to v:

$$\delta_v = \Psi\big(\rho(v, V_i), g(V_i)\big) \tag{2}$$

where g is a function of V_i, and Ψ is the redistribution function. In Ψ, $\rho(v,V_i)$ is usually used as a multiplicative coefficient of $g(V_i)$. In the following applications, function Ψ is defined for voxels V_i located in a neighborhood N of v, which can be defined as the set of voxels V having a non-null intersection with v ($v \cap V \neq \varnothing$ – see fig 5). Moreover, $\Phi(v)$ can also be processed from values $\varphi_{v'}$ where $v' \in N'(v)$, neighborhood of v at the higher resolution.

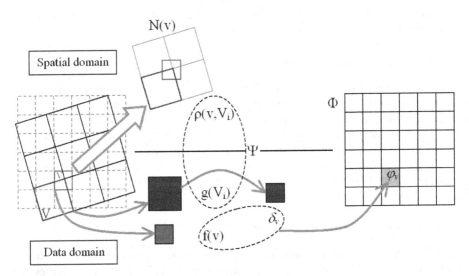

Fig. 5. In the final image Φ (result of the fusion), the activity φ_v of voxel v is processed from both information f(v) in the high-resolution image, and information $g(V_i)$ held by voxels V_i in the low-resolution image, V_i being the voxels which belong to the neighborhood N(v) of v. Information $g(V_i)$ have to be injected into v (δ_v) by the redistribution function Ψ using spatial coefficients $\rho(v,V_i)$, which express spatial relations between v and V_i.

The choice of both spatial coefficients and processing operator have been guided by the digital nature of image information. Indeed, the question of studying spatial relations between two discrete lattices is very close to the classical discrete coordinates changing problem. Discrete and computational geometries provide efficient tools to answer such a problem.

3   Discrete Approach of the Multiresolution Problem

The choice of the voxel model is of high importance, since it represents the influence region of a numerical activity during the measurement, when going back to the physical meaning of the image. In 2D images, pixels are usually shown as square or even rectangular areas, filled by a uniform gray level (or color) depending on the numerical activity. An hexagonal model is for instance a better representation of a round influence area (*e.g.* orthographic projection of a spherical region) than a square one. Nevertheless, for simplification purposes, only the case of cubic voxels is addressed here. Discrete geometrical tools used in the following have been chosen on this assumption.

3.1   Digital Coordinates Changing Problem

Whatever the chosen geometrical model of the pixels or voxels [5], the extension of the tilling they create to the whole space may be considered as a discrete coordinate system. Spatially matching one of the images with another thus boils down to the discrete coordinates changing problem (*i.e.* how to accurately express, in a base B_2, discontinuous spatial information from a base B_1, $B_1 \neq B_2$). A solution was proposed in [13] with a suitable formalism for 3D image fusion. When the images have close spatial resolutions, the ratio between edge length of the low and high-size voxels is not sufficient to disregard the committed error when rounding to integers the results of classical base changing formulas. In this case, one shall determine the volume of the geometrical intersection between unit volumes of the grids and use this value in the compulsory interpolation step.

Using the intersection volume between voxels v and V as spatial coefficients $\rho(v, V)$, formula (2) thus becomes:

$$\delta_v = \sum_{V \cap v} \rho(v, V) g(V) \tag{3}$$

where $V \cap v$ refers to the neighborhood of v presented in section 2.3. Formula (3) processes information δ_v associated to v as a combination of information $g(V)$ associated with surrounding voxels V, where the contribution of V is proportional to its common volume with v. This calculation mode is close to a linear interpolation of $g(V)$, but uses the maximum of information available in the image structure, thus minimizing the required hypothesis on the distribution of the measured physical phenomenon.

When building image models, the high-resolution image H is chosen as geometrical reference. Its voxels v are thus represented as unit cubes. Therefore, spatial coefficients $\rho(v, V)$ have values ranging from 0 to 1, and

$$\sum_{V \cap v} \rho(v, V) = 1 \tag{4}$$

Each piece of information δ_v thus belongs to the same value range as original information $g(V)$ associated with voxels V.

3.2 Building of the Geometrical Model

Voxel sets of both images are modeled by cubic tilings of the space. Cubic models of the high-resolution image H have unit lengths and are associated with the canonical base of the image space. Voxel models of the low-resolution image L are cubes in general position (see fig. 2). Let **T** be a registration transform aligning L data on H data (**T** is of course not a applied to image data, but stems from a classical registration processes). Since only rigid transforms are addressed here, **T** is composed of a rotation and a translation (6 parameters). The difference in spatial resolution between acquisition systems implies a third part in the transform, based on the application of a scaling factor. As accurately described in [3], the generator vectors of cubes in general position are the images of canonical unit vectors by **T**, and all cubes of the tiling are obtained by translation of the origin voxel.

3.3 Processing of Spatial Coefficients

The intersection volumes between high and low-size voxels are polyhedral (see fig. 6.a), and are computed using an efficient cube intersection algorithm. The processing cost of a single intersection volume is lower than using a general algorithm for the intersection of convex polyhedrons [14], thanks to the use of analytic formulas linked with inherent cube symmetries (see fig. 6.b).

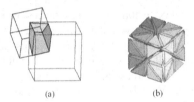

(a)   (b)

Fig. 6. Example of intersection volume (a) processed thanks to the octahedral group of cube symmetries (b)

The principle is to run through the 6 faces of both cubes, processing a polygonal bound of the polyhedral volume at each iteration (at most 12 faces). Analytic formulas provide the coordinates of candidate points for vertices of a given face, as a function of its distance to the central point of the reference cube (low-size voxel). To do so, the normal vector of the face is first considered as constrained in a canonical cone of the space (see fig. 6.b), and real points are then retrieved thanks to cube symmetries. Finally, the set of candidate points is completed, and vertices are selected thanks to a polygon intersection algorithm. For more details about the intersection algorithm, see [13] and [3].

4   Modeling of 3D Images

The intersection algorithm proposed in [13] builds an intersection polyhedron between two given voxels in linear time (depending on the dimension of input objects and the maximum number of vertices of its faces; both are considered as constant is this paper). The volume itself is also processed in linear time (sum of pyramidal volumes). However, the whole processing time remains prohibitive in the case of large 3D image fusion (*e.g.* 256^3 voxels). Indeed, for each voxel of the high-resolution image, intersection volumes are processed in a 3×3 neighborhood of voxels V in order to verify condition (4). Thus, we also propose an improvement of the image geometrical model, using properties of discrete geometrical objects and rational numbers to reduce processing regions: when cube generating lines obtained from transform **T** are approached by integer directions (lines with rational slopes), some intersection periods appear between discrete objects of the models.

4.1 Properties of Discrete Objects

Let v_1 be a vector with integer coordinates (a b) and D be the Euclidian line generated by v_1. All the points k(a b), $k \in Z$, belong to D. The set of pixels covered by D forms a discrete line $\mathcal{D}$ (see fig. 7.a). Each pixel based on the point k(a b), $k \in Z$, belongs to $\mathcal{D}$, which is thus composed by the repetition of a pixel pattern (pixels of $\mathcal{D}$ located between points k(a b) and (k+1)(a b), $k \in Z$). The period of the pattern occurrence is a function of the coordinates of v_1 and thus of α, rational slope of D. Such a property is preserved with a generating vector v'_1 which coordinates are real numbers with a rational ratio α.

If α is an irrational number, the period becomes infinite. In this case, many solution have been proposed to approximate the generating vector with an integer vector. Most are based on the development of α in continued fraction [15]. Such approximations can also be used to reduce the complexity in the case of large periods.

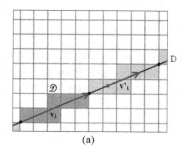
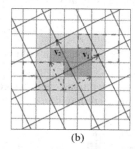

(a)	(b)

Fig. 7. Periodicity of discrete structures. In 1D (a), the discrete line $\mathcal{D}$ is built by reporting the dark-gray pattern. The only condition is that line D have a rational slope (rational generating vectors v_1 or v'_1). In 2D (b), the intersection between the canonical grid and the lattice generated by integer vectors v_1 and v_2 is also periodic. The dark-gray pattern is deduced from v_1 and v_2, and the light-gray one is obtained from vertices having the same abscissa/ordinate when reporting the dark-gray pattern.

In the 2D case, the intersection of two grids is also periodic if generating vectors are either integer or rational (see fig. 7.b). In the 3D case, things are more complicated, since cubes in general position don't have integer vertices. However, when the periodicity exists, it can be reduced by studying multidimensional continued fractions.

Real 3D periods will thus never appear in the intersection of voxel models, whatever the generating vectors of high-size voxels. Despite this problem, reducing the processing time of the intersection algorithm remains possible thanks to properties of *e.g.* 2D involved objects. For instance, such objects appear at the first step of the algorithm: analytic formulas used in this algorithm aim at processing the intersection the intersection between a cube and the support plan of the searched face. Since the intersection of a plan with a cubic tilling of the space is doubly periodic, the processing area can be restricted to a portion a each support plan. Moreover, support plans of the searched faces being organized into groups of parallel objects, the restricted processing area is common to many plans in the image structures.

4.2 Changes in the Geometrical Model

In order to introduce such periodicity properties into the geometrical model, generating vectors of voxels from the low-resolution image have to be approached by vectors with integer directions (diophantine approximations). The first directions is simply approached by an iterative research of the closest integer point. Approximations of other directions are constrained by the orthogonal nature of generating lines to obtain cubic models. The second approximation is thus processed using a geometric version of the Jacobi-Perron algorithm [16]. The third one is simply obtained by cross-product of the two other vectors, since this operator preserves integer/rational features of the objects.

4.3 Influence of the New Model on Intersection Volumes

To assess the influence of these model approximations on the processed intersection volumes, figure 8 presents relative errors committed on 8 volumes, supposed to be the most affected ones. The 8 voxels are those associated with vertices of a cubic volume of 256^3 voxels. The relative error is processed as follows:

$$E_{v \cap V} = (\rho_{app} - \rho_{init})/(\rho_{init}) \tag{5}$$

where ρ_{app} is the intersection volume processed after approximation of the concerned direction, and ρ_{init} is the intersection volume processed from the original image model. For each v among the 8 tested voxels, a unique V is chosen as the one having the largest intersection with v.

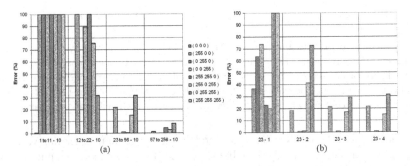

Fig. 8. Relative error committed on intersection volumes, when approximating the first (a) and the second (b) directions for 8 voxels. In part (a), the number of iteration of the approximation algorithm varies from 1 to 256, while this number is fixed to 10 for the second approximation algorithm. In part (b), 23 iterations are performed for direction one.

When running the first approximation algorithm, the closest integer point only change 3 times on the chosen voxel set. The first selected approximation is vector (1 0 0). This rough approximation has only a few influence on the origin voxel, while the intersection volume is null for all other voxels. When the number of iterations increases, voxels located in a plan orthogonal to the approached direction are les affected than other voxels. Most errors remain higher than 30% with less than 23 iterations. The best approximation is reached with 57 iterations (only $\approx 1/5$ width of the voxel set), where all errors are lower than 10%. The first committed error is lower in the case of the second approximation, since the integer coordinates are searched within a constrained set of points, resulting from the first approximation. The Jacobi-Perron algorithm quickly converges, but the norm of resulting vectors also quickly increases, thus implying large periods for discrete object properties. These results show that the image model have to be built as a compromise between processing efficiency and accuracy.

5 Application to the Fusion of Multimodal Medical Images

Two applications of the fusion process by activity redistribution were tested in the field of medical image analysis, using voxel intersection volumes as spatial coefficients. In each case, the image fusion is performed as an implementation of the multimodality approach of medical imaging for the help in diagnosis of brain pathologies such as Alzheimer type dementia and Parkinsonian syndromes.

The image fusion is performed between an MR image (anatomical information) as high-resolution data set, and a SPECT image (functional information, either brain perfusion or neurotransmission) as low-resolution data-set.

5.1 Synthesis of a Support Image for New Diagnostic Information

The preliminary study about image synthesis is presented in [8]. Image fusion between MR and SPECT data comes out to process fuzzy maps for brain tissues (cerebrospinal fluid (CSF), white matter (WM), gray matter (GM) and hypoperfused gray matter), and then finally to present all the membership information on an unique synthetic image. In this final fusion result, each membership was only associated with the mean gray level of the related class. A local information is now introduced in image Φ at the voxel level by activity redistribution (see fig. 9). In this case, formula (1) becomes:

$$\varphi_v = \left(\sum_C \pi_C(v)\mu_C + \delta_v \right) \bigg/ \left(\sum_C \pi_C(v) + \sum_{V \cap v} \rho(v, V) \right) \tag{6}$$

where $\pi_C(v)$ is the membership of voxel v to the tissue class C, μ_C is the mean value of the functional activity for class C. Value δ_v is processed from the spatial coefficient $\rho(v,V)$ as in formula (3), with $g(V) = L(V)$, the original activity of V in the SPECT image (ρ simply weights the part of activity L(V) injected into voxel v – see formula (7)).

The process has been applied to a brain perfusion SPECT ([^{99m}Tc]-ECD) image (see fig. 10). The anatomical MR image was T_1 weight with 0.94 mm large voxels (size in the slice plane). The SPECT images is isotropic, with a voxel size of 2.33 mm. The slice thickness of the MR image is 1.5 mm, but the used registration process (coming out to transform **T**) is based on an interpolated version of these data to cubic voxels.

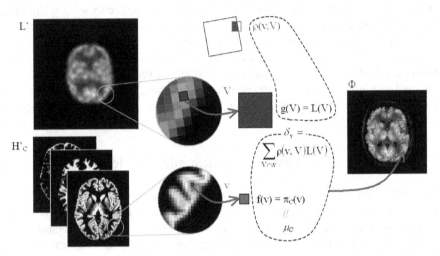

Fig. 9. In this visual application of the multiresolution fusion, high-resolution anatomical information are injected into voxel v of the final image Φ through v memberships to tissue classes C (high-resolution membership maps H'$_C$), weighting the mean functional activity levels μ_C. The resulting quantity is homogenous to the other kind of information, and has a high visual representation potential. These values are mixed with the low-resolution functional information (shown as the registered but not interpolated image L') using formulas (1)/(2) where g(V) is the functional activity L(V), and using intersection volumes as spatial coefficients ρ(v,V) (formulas (6)/(7)).

Perfusion images refer to the activity of neurons in the whole brain. The corresponding functional activity is thus a global information, with low frequencies components. For the help in diagnosis, it has to be emphasized in the region of cerebral cortex. The concentration level of the tracer within CSF structures is null. Variations are visible in the WM and mostly in the GM (activity ratio estimated to 1/4). High activity levels in the cortical region are clearly visible, since the mean activity of this structure is already high. Nevertheless, hypoperfusion zones are mostly visible in regions with low base activity, *i.e.* white matter. Visible results are consistent with expected behavior of the SPECT tracer.

This first application allows a qualitative assessment of the proposed method. Even in the case of the image synthesis process, a quantitative analysis consists in the assessment of information conservation in a given region of interest (ROI). The following application gives elements to answer this question.

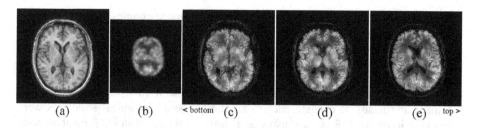

Fig. 10. Axial slices of the synthetic image (c-e) obtained by fusion of MR (a) and brain perfusion SPECT (b) images

5.2 Activity Quantitation in Anatomical ROIs

Most numerical indexes extracted from image data in the context of medical image analysis are based on the measurement of mean image activities within anatomical structures of interest, linked with the explored pathology. The fixation rate represents a relative variation with respect to a reference activity (*e.g.* mean activity in a ROI where the pathology has only a few effects), in order to compare inter-patient results and allow an automated classification of the patients.

In such an application, the model of anatomical information simply consists in binary maps representing the segmented structures of interest. Using the redistribution principle, the activity assigned to a voxel v is simply equal to the redistributed activity δ_v, processed as in formula (3) with $g(V) = L(V)$, the original activity of V in the SPECT image:

$$\delta_v = \sum_{V \cap v} \rho(v, V) L(V) \tag{7}$$

The final fusion information Φ is the fixation rate of the SPECT tracer in each structure of interest s, which is a function of the mean activities μ_s within s, instead of single voxel values φ_v. Let S be the binary map of a given structure of interest s (either specific to the pathology or non-specific). The mean value within structure s can be processed by the following formula

$$\mu_s = \left(\sum_v S(v) \delta_v \right) \Big/ \left(\sum_v S(v) \right) \tag{8}$$

which is a classical arithmetic average formula, and correspond to a modified form of equation (1) where function f is the binary map S, equal to 1 when voxel v belongs to the structure of interest, else equal to 0. Moreover, formula (8) is a multiplicative fusion operator instead of being additive as formula (1).

Results obtained from activity quantification in a numerical phantom of dopaminergic neurotransmission (parkinsonian syndromes) are presented in [2]. A relative measurement error is processed (with respect to an *a priori* mean activity level) for different selections of slices in a binary map of putamens, in order assess the quantification method on images suffering from partial volume effect resulting of the multiresolution aspect of data (shown on fig. 1).

The quantification process was also applied to 14 image couples (T_1 weighted MR image/dopaminergic neurotransmission SPECT) in the context of the differential diagnosis of parkinsonian syndromes. Among the 14 subjects, 5 are cases of idiopathic Parkinson's disease (IPD), 5 are multiple system atrophy (MSA), and 4 are affected by one of the preceding parkinsonian syndromes, but have an uncertain diagnosis. Brain anatomical structures involved in such illnesses are mainly the putamens and heads of caudate nuclei (striatum). These anatomical structures were segmented in the 14 MR images using an algorithm presented in [17]. Fixation rates were processed for both left and right-hand structures and some substructures (anterior and posterior parts of the putamens), using a part of the brain occipital lob as non-specific region. In order to assess the proposed method (denoted method 1), all the quantitative indexes were also processed by direct measurement in the registered SPECT image (which implies a data interpolation, denoted method 2). One half of the patients were classified using these numerical indexes, by a discriminant analysis (7 cases of known diagnosis patients being used for the learning stage). Results of this automatic analysis are presented in fig. 11.

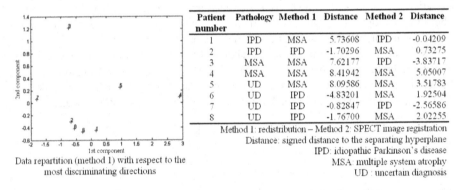

Data repartition (method 1) with respect to the most discriminating directions

Patient number	Pathology	Method 1	Distance	Method 2	Distance
1	IPD	MSA	5.73608	IPD	-0.04209
2	IPD	IPD	-1.70296	MSA	0.73275
3	MSA	MSA	7.62177	IPD	-3.83717
4	MSA	MSA	8.41942	MSA	5.05007
5	UD	MSA	8.09586	MSA	3.51783
6	UD	IPD	-4.83201	MSA	1.92504
7	UD	IPD	-0.82847	IPD	-2.56586
8	UD	IPD	-1.76700	MSA	2.02255

Method 1: redistribution – Method 2: SPECT image registration
Distance: signed distance to the separating hyperplane
IPD: idiopathic Parkinson's disease
MSA: multiple system atrophy
UD : uncertain diagnosis

Fig. 11. Results of the discriminant analysis for methods 1 and 2

Since this analysis is only performed on a few data, it remains impossible to draw a conclusion about the method efficiency, even if such a parametric test remains valid [18]. However, this test shown that the use of the proposed fusion method is possible to quantify the activity in biomedical images. Only on subject was misclassified with method 1 (patient 1), against 2 subjects using method 2. Moreover, this subject (1) has coherent features with those of his real class (distance to the hyperplane). In a latter assessment protocol, the case of patients with an uncertain diagnosis was studied again, and the results confirm the classification proposed by method 1.

6   Conclusion

In this paper, a new method is presented for multiresolution image fusion. This method is based on a new fusion scheme where image numerical activities are processed apart from image spatial features. The aim is to preserve original information until the final fusion step, thus avoiding *e.g.* rounding errors. Moreover, the major interests of this method (by comparison with a data interpolation) are: the ability to introduce the original image information into complex fusion operators, and the maximum use of available information to avoid arbitrary hypotheses, *e.g.* on the spatial distribution of image activities.

Spatial features of the images to fuse are managed through geometrical models. In the spatial domain, a fusion is performed between these models. The result is thus mixed with the fusion operator applied in the data domain to obtain the final result. After the fusion in the spatial domain, spatial relations between both images are represented by a simple set of numerical values, thereafter used as weighting coefficients for the data domain information. A geometrical model has been proposed, based on the representation of image voxels, leading to use intersection volumes between these voxels as spatial coefficient for the redistribution of image activities.

The geometrical model and the spatial fusion have been implemented in the case of images with cubic voxels, *i.e.* isotropic images. But the fusion scheme remains valid for other kind of images (non-isotropic images, non-cubic models of voxels, *etc.*). The implementation of the method in other cases involves both the redefinition of the geometrical model and the conception of an efficient intersection algorithm adapted to the shape of voxel models. Even with the proposed geometrical model, the efficiency of the algorithm may be improved by introducing the management of intersection periods.

References

1. Barra, V., Boire, J.Y.: A general framework for the fusion of anatomical and functional medical images. NeuroImage 13 (2001) 410–424
2. Montagner, J., Barra, V., Reveillès, J.P., Boire, J.Y.: Multiresolution images fusion for the quantification of neuronal activity: a discrete approach. Proceedings of the 3rd IASTED International Conference on Biomedical Engineering, Innsbruck-Austria (2005) 1–6
3. Montagner, J., Barra, V., Boire, J.Y.: Modeling of a multimodal image aggregation process using discrete geometry. Proceedings of the 1st Open International Conference on Modeling and Simulation, Clermont Ferrand-France (2005) 431–439
4. Bloch, I., Géraud, T., Maître, H.: Representation and fusion of heterogeneous fuzzy information in the 3D space for model-based structural recognition – Application to 3D brain imaging. Artificial Intelligence 148 (2003) 141–175
5. Chassery, J.M., Montanvert, A.: Géométrie discrète en analyse d'images. Hermès, Paris-France (1991) (in French)
6. Bloch, I.: Information combination operators for data fusion: a comparative review with classification. IEEE Transactions on Systems, Man, and Cybernetics 1 (1996) 52–67
7. Zadeh, L.A.: Fuzzy sets as a basis for a theory of possibility. Fuzzy Sets and Systems 1 (1978) 3–28
8. Colin, A., Boire, J.Y.: MRI-SPECT fusion for the synthesis of high resolution 3D functional brain images: a preliminary study. Computer Methods and Programs in Biomedicine 60 (1999) 107–116
9. Shafer, G.: A mathematical theory of evidence. Princeton University Press, Princeton-USA (1976)
10. Pajares, G., De La Cruz, J.M.: A wavelet-based image fusion tutorial. Pattern Recognition 37 (2004) 1855–1872
11. Matsopoulos, G.K., Marshall, S., Brunt, J.N.H.: Multiresolution morphological fusion of MR and CT images of the human brain. Proceedings of the IEE Vision, Image and Signal Processing Conference (1994) 141(3) 137–142

12. Soret, M., Koulibaly, P.M., Darcourt, J., Hapdey, S., Buvat, I.: Quantitative accuracy of dopaminergic neurotransmission imaging with 123I SPECT. Journal of Nuclear Medicine 44 (2003) 1184–1193
13. Reveillès, J.P.: The geometry of the intersection of voxels spaces. Electronic Notes in Theoretical Computer Science 46 (2001)
14. Hertel, S., Mäntylä, K., Mehlhorn, K., Nievergelt, J.: Space sweep solves intersection of convex polyhedra. Acta Informatica 21 (1984) 501–519
15. Hardcastle, D.M., Khanin, K.: Continued fractions and the d-dimensional Gauss transformation. Communications in Mathematical Physics 215 (2001) 487–515
16. Moussafir, J.O.: Voiles et polyèdres de Klein. Géométrie, algorithmes et statistiques. Thesis, University Paris IX-Dauphine, Paris-France (1992) (in French)
17. Barra, V., Boire, J.Y.: Automatic segmentation of subcortical brain structures in MR images using information fusion, IEEE Transactions on Medical Imaging 20(7) (2001) 549–558.
18. Romeder, J.M.: Méthodes et programmes d'analyse discriminante. Dunod, Paris, France, 1973 (in French)

Analysis and Visualization of Images Overlapping: Automated Versus Expert Anatomical Mapping in Deep Brain Stimulation Targeting

Lemlih Ouchchane[1,2], Alice Villéger[2], Jean-Jacques Lemaire[2,3],
Jacques Demongeot[4], and Jean-Yves Boire[1,2]

[1] Centre Hospitalier Universitaire de Clermont-Ferrand,
Département de Santé Publique, Unité de Biostatistiques,
Télématique et Traitement d'Images, 63000 Clermont-Ferrand, France
{lemlih.ouchchane, j-yves.boire}@u-clermont1.fr
[2] Université d'Auvergne, Equipe de Recherche en Imagerie Médicale,
ERIM-EA 3295 ERI 14 ESPRI-INSERM, 63000 Clermont-Ferrand, France
alice.villeger@u-clermont1.fr
[3] Centre Hospitalier Universitaire de Clermont-Ferrand,
Service de Neurochirurgie A, 63000 Clermont-Ferrand, France
jjlemaire@chu-clermontferrand.fr
[4] Université Joseph Fourier, Laboratoire Techniques de l'Imagerie, de la Modélisation
et de la Cognition, TIMC-IMAG UMR CNRS 5525, 38706 Grenoble, France
jacques.demongeot@imag.fr

Abstract. In surgical treatment of Parkinson's disease, deep brain stimulation requires high-precision positioning of electrodes, needing accurate localization and outlines of anatomical targets. Manual procedure of anatomical structures outlining on magnetic resonance images (MRI) takes about several hours. We proposed an automated localizing procedure aiming to shorten this task to some seconds. Different parameters were simultaneously assessed in our algorithm undertaking segmentation of anatomical structures. Intraclass correlation coefficients (ICCs) were computed for centers of gravity coordinates of structures between manual expert-mapped MRI and automated-mapped MRI. Tanimoto coefficients were computed accounting for pixels overlapping between these two procedures. Although ICCs showed almost perfect concordance, TC provided further information with a quite severe value about 35%. For both criteria, results were variable regarding each parameter in our process. With such complex results to relate, their presentations were enhanced using visualization methods resembling that of the generalized Case View method.

1 Introduction

Interests of deep brain stimulation in Parkinson's disease (PD) treatment are well-established [1][2]. This surgical technique requires high-precision positioning of electrodes into specific deep brain structures [3], especially when direct visualization of anatomical target is carried-out on magnetic resonance images (MRI) [4][5]. Expert mapping on MRI is very long-standing. This procedure, carried-out almost manually

P.P. Lévy et al. (Eds.): VIEW 2006, LNCS 4370, pp. 137–151, 2007.
© Springer-Verlag Berlin Heidelberg 2007

by a neurosurgeon accustomed to numerical image issues, is taking about several hours [4]. Therefore automated methods are useful to shorten this essential task.

An algorithm of automated anatomical structures mapping, undertaking segmentation of specific anatomical structures, is developed in our laboratory. It is providing both topography and label of structures on MRI and reduces the duration of this procedure to some seconds. However, this algorithm needs assessment regarding its accuracy to localize deep brain structures and also regarding its validity to correctly label every pixel in MRI according to structure membership [6][7]. The aim of this study is to direct choices in options that may lead to enhance this procedure. We assessed different parameters in our algorithm basing on comparison between these two procedures (i.e. expert versus automated mapping) assessing concordance on centers of gravity coordinates of structures (SCG), abscissa and ordinate, and computing Tanimoto coefficients (TCs). The latter might prove to be particularly relevant because providing a more severe assessment of global similarity for each observation and accounting for pixels overlapping between images resulting from these two procedures. Since different parameters are assessed simultaneously in this study, results are multi-factorial and thus complex to recover. Their presentations was carried-out using visualization method to comprehend them in a synoptic view [8][9].

2   Patients, Materials and Methods

2.1   Patients

Ten patients suffering from PD, in whom deep brain surgery was indicated, were enrolled in this study. Deep brain structure targeting was carried-out on MRI acquired in stereotactic conditions before planning surgical path and positioning of electrodes [5].

2.2   Tissue Characterization

MRI slices are composed of $512*512=262\ 144$ pixels with 256 gray shades (8 bits).

First, each MRI was submitted to a tissue characterization that generated three fuzzy membership maps belonging to each tissue class, i.e. white matter, gray matter and cerebrospinal fluid [8]. A fourth class, referred to as "background", was not accounted for in the remainder of the process. A fuzzy membership map is an image mapping each pixel to a value in [0;1]. This value is a membership degree, called μ in the remainder. It quantifies how much a pixel seems to belong to a particular tissue class. Figure 1 represents membership as gray levels: white pixels ($\mu=1$) surely belong to a class, while black pixels ($\mu=0$) surely do not belong to that class, and intermediate gray pixels account for uncertainty on membership degrees. The last picture illustrates a fusion of all four tissue maps: cerebrospinal fluid appears as white, gray matter as light gray, white matter as dark gray, and background as black (such color mapping reproduces usual contrasts observed in MRI). This image has been "defuzzified", i.e. pixels are categorically assigned to the tissue class for which their membership degree is the highest (Fig. 1, see also Fig. 5).

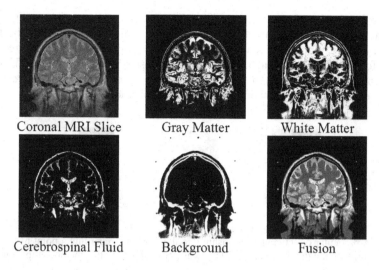

Fig. 1. Tissue characterization of a coronal MRI slice (T2 weighted)[1] with gray and white matters, cerebrospinal fluid, background membership maps and the fusion of all tissue classes

A restricted and standardized rectangular region of interest was defined for each patient's MRI that included the 6 main deep brain structures involved in our topic; for white matter: pyramidal and mamillothalamic tracts grouped as a unique structure (PT), for gray matter: caudate nucleus (CN), thalamus (TH), and subthalamic nucleus grouped with substantia nigra as a unique structure (SN), and lastly for cerebrospinal fluid: the third ventricle (TV) and the lateral ventricle (LV).

2.3 Automated Procedure of Anatomical Structures Mapping

The procedure uses fuzzy sets as a common modeling framework to represent every sources of information. Tissue information results from the classification performed on the patient's MRI. Fuzzy maps are used to represent prior expert knowledge about spatial relationships between structures. Focusing on relative position in every pair of structures, this information is stored inside of an anatomical model consisting in a graph where vertex represent structures, and edges represent directional relationships [6][7].

We designed a process fusing both sources of information (i.e. patient's MRI and theoretical model), leading to the localization of structure. Each structure map is initialized with the fuzzy map of the tissue it belongs to. The process then folds into two steps. First, structures belonging to the same tissue class are separated from one another, propagating directional relationship from the most central structure to the most outer one, then back from border to center. The second step consists in an iterative refinement of that first solution, propagating every relationship between all structures in the model; until convergence towards a stable solution is achieved with every directional constraint defined in the model respected. The resulting membership map

---

[1] Source: J.-J. Lemaire, Univ. Hosp. of Clermont-Ferrand, Neurosurgery Department, France.

obtained for each structure can then be interpreted as a segmentation result. This procedure relies heavily on directional relationships, which are propagated by fusing a directional fuzzy map to a structures membership map. However, there are different ways for modeling a particular directional relationship into a fuzzy map. Three different fuzzy models (referred to as "methods") were used; respectively severe (method=1), indulgent (method=2), and cautious (method=3) (Fig. 2).

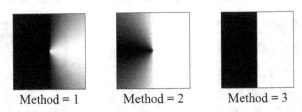

Method = 1   Method = 2   Method = 3

Fig. 2. A particular directional relationship ("to the right of", regarding a point in the center of the square) with the three different models (white corresponding to a membership value of 1)

We also experimented with an enhanced version of the cautious method, adding a supplementary step at the end of the original process that researches connected components (method=4). An additional parameter is also studied: the defuzzification threshold (referred to as "threshold"), which could either be applied (threshold=1), or not (threshold=0), at the very end of the process. This defuzzification step consists in mapping every pixel either to 0 or 1, depending on whether its membership degree is lower or greater than a specific threshold (set to 0.5 in our study) [6][7].

2.4 Assessments of the Automated Mapping Procedure

SCG were localized with abscissa and ordinate (abscissas axis on the upper edge of MRI and ordinates axis on inter-hemispheric line). SCG abscissa (1) and ordinate (2) are computed after the fuzzy membership map is obtained for each structure.

$$\text{SCG abscissa} = x_{SCG} = \frac{\Sigma(x*\mu(x,y))}{\Sigma(\mu(x,y))} \quad . \tag{1}$$

$$\text{SCG ordinate} = y_{SCG} = \frac{\Sigma(y*\mu(x,y))}{\Sigma(\mu(x,y))} \quad . \tag{2}$$

Where (x,y) are coordinates of points weighted with its membership degree ($\mu(x,y)$) regarding a particular structure.

ICCs between SCG coordinates were computed to assess accuracy of the automated procedure in localizing anatomical structures [11][12][13]. ICCs were used aiming to detail trends between options or anatomical structures, sorting options from low to high accuracy on SCG coordinates localization. Thus, no statistical inference was carried-out on ICCs between methods or structures.

TCs were computed to assess global similarity between expert-mapping and automated-mapping MRI. TCs account for pixels information in the whole MRI and not only over centers of gravity or outlines of structures [14]. MRI is regarded as X by Y matrix with binary information, 1 when structure includes pixel, 0 otherwise.

Considering superimposition of both images, respectively expert (used as reference, appearing as first value in the following binary couple) and automated, 4 pixel types are defined: {0,0} referred to as true negative (TN), {1,1} true positive (TP), {1,0} false negative (FN) and {0,1} false positive (FP). TC upon structure is defined as ratio of intersection to union of 1-coded pixels (3), with a possible range from 0 to 1 (or 0% to 100%), TC upon background is defined on 0-coded pixels (4), with maximum equal 1 (Fig. 3).

$$TC \ (structure) = \frac{TP}{TP+FN+FP} \ . \tag{3}$$

$$TC \ (background) = \frac{TN}{TN+FN+FP} \ . \tag{4}$$

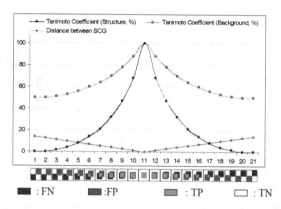

Fig. 3. Range of Tanimoto coefficient upon structure (TC upon background and distance between SCG) for 21 different overlapping configurations between simulated square images

Since each of our 10 patients was assessed for each of the 6*4*2=48 combinations of modalities of respectively structure/method/threshold, analyzes of parameters effects were carried-out in a pure within-subject design [15] (i.e. in all of our 10 patients, for each method and each threshold condition, each structure was localized and labeled using the two procedures (i.e. automated versus expert). When analyzing the whole data set, effects of automated procedure methods and structures on TC and possible interactions were tested on SAS [16] using a linear mixed model accounting for repeated measures in the same patient [17]. Restricted analyzes for single factor effects were carried-out and we computed, when appropriate, post-hoc multiple comparisons using Tukey's honestly significant difference test [18].

2.5 Visualizing Parameters Effects of the Automated Mapping Procedure

Results on ICCs or TCs can be displayed in a table e.g. with a row for each method, a column for each structure, splitting table in two parts whether a defuzzification threshold is applied or not, each cell containing the value of ICC or TC of a particular combination of these parameters. Assuming method as an ordinal variable (with some refinement in methods capability), structure as nominal, threshold as binary and ICCs or TCs as continuous, results can generate an image hugely inspired from PP Lévy *et al* works on methods for visualizing complex data such as the case mix [8][9]. Keeping the same presentation that forms the table aforementioned, we proposed to replace value in each cell by a color coding regarding values of ICC or TC. Color coding is beginning with dark blue for lowest values, ranging through shades of blue, then cyan, green, yellow until red, and ending with dark red for highest values. This process was carried out on MATLAB [19] which result is an image where each cell appears like a pixel with a particular color regarding the ICC or TC value it contains in the original table.

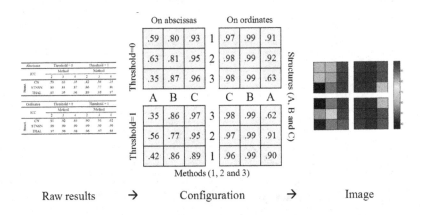

Raw results   →   Configuration   →   Image

Fig. 4. Visualization of multifactorial results: No. 1, 2 and 3 refer to the method assessed which is assumed to be ordinal, A, B and C refer to structures and the last binary variable split the image in two parts (threshold = 0 versus threshold = 1). Visualization is possible on ICCs for both abscissas and ordinates with images placed symmetrically regarding structures

While this view is synoptic, to keep it easily interpretable, we presented the image with symmetry axes, with methods disposed in reverse order whether pixels are on the upper-side (Threshold = 0) or on the lower-side of the image (Threshold = 1). Since ICCs are computed for both abscissas and ordinates, two images are displayed symmetrically regarding structures too (Fig. 4).

For TC, results are displayed with the same method but more simply since there is no splitting regarding coordinates. We pursued the method, visualizing results of Tukey's test focusing on vertical or horizontal strips of the image. The result is a color bar displayed beside, above or below the original image regarding respectively method or structure simple effect. A symbol for insignificant differences was also displayed using common underscores.

3   Results

3.1   Expert-Mapping Versus Automated-Mapping MRI

Expert-mapping MRI is considered as reference information and the SCG coordinates resulting from this MRI are referred to as "expert localization" or "expert mapping" in contrast with "automated localization" that refers to the "automated-mapping" MRI (Fig. 5).

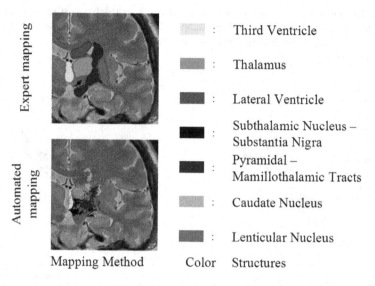

Mapping Method   Color   Structures

Fig. 5. Coronal slice MRI with expert and automated-mapping of anatomical structures

SCG are displayed for each procedure in Fig. 6, showing a notable accuracy for the automated procedure in localizing third ventricle, lateral ventricle, subthalamic nucleus-substantia nigra and thalamus. On the contrary, lenticular nucleus and pyramidal-mamillothalamic tracts ordinates appeared biased, similarly to caudate nucleus which appeared biased for both abscissa and ordinate (see also Fig. 5). These trends are emphasized on scatter plots for both procedures of SCG localization, where caudate nucleus and pyramidal-mamillothalamic tracts appeared clearly moving away from identity line (Fig. 7).

Figure 6 also displays an example of TC computation, here on thalamus with a particular distribution of pixel types as described in section 2.4. True positive are green colored, false negative are blue colored, while false positive, red colored are almost absent for this particular mapping (true negative were not displayed here). In this example, TC equal 56.4 (%) on thalamus using method 4 and applying a defuzzification threshold.

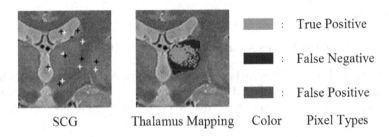

SCG   Thalamus Mapping   Color   Pixel Types

Fig. 6. Structures center of gravity (SCG) coordinates: Expert-labeled SCG (in black) versus automated-labeled SCG (in white) and an example of TC calculation on thalamus mapping. TC computation on thalamus with true positive (green), false negative (blue) and false positive (red) pixels type.

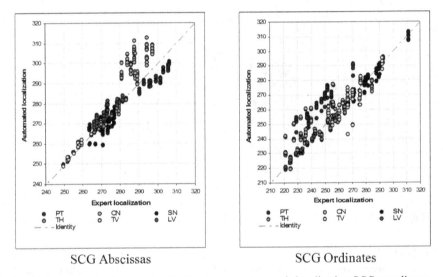

SCG Abscissas   SCG Ordinates

Fig. 7. Scatter plots of expert-localization versus automated- localization SCG coordinates

3.2 ICCs on SCG Coordinates

Mean value of ICC on the entire dataset was 0.89. Computed separately on SCG abscissas and ordinates, average ICCs were respectively 0.86 and 0.93. Assessing methods capability in localizing structures, mean ICC (and between parentheses, the same value applying a defuzzification threshold) ranged as follows: on SCG abscissas, 0.87 for methods 1 (0.86) and 3 (0.84), 0.86 for method 2 (0.85) and 0.85 for method 4 (0.85), while on SCG ordinates, 0.96 for methods 3 (0.96), 0.95 for method 1 (0.95) and 2 (0.95), and 0.84 for method 4 (0.83).

Assessing localization efficiency regarding each structure, average ICC ranged as follows: on SCG abscissas, 0.98 for third ventricle (0.98), 0.95 for lateral ventricle (0.93) and thalamus (0.94), 0.91 for pyramidal-mamillothalamic tracts (0.92), 0.81 for subthalamic nucleus-substantia nigra (0.81) and 0.57 for caudate nucleus (0.50), while on SCG ordinates, 0.99 for lateral ventricle (0.99) and subthalamic nucleus-substantia nigra (0.99), 0.98 for thalamus (0.97), 0.92 for pyramidal-mamillothalamic tracts (0.91), 0.85 for caudate nucleus (0.84) and 0.83 for third ventricle (0.83).

Detailed results are displayed in Table 1 and Table 2.

Table 1. ICCs for abscissas between expert and automated localization of SCG displayed for each combination of method/threshold and structure. Structures are pyramidal-mamillothalamic tracts (PT), caudate nucleus (CN), subthalamic nucleus-substantia nigra (SN), thalamus (TH), third ventricle (TV) and lateral ventricle (LV).

ICCs on abscissas (No. = 10)	Threshold = 0 Methods				Threshold = 1 Methods			
	1	2	3	4	1	2	3	4
PT	.91	.89	.90	.93	.91	.98	.84	.93
CN	.67	.59	.63	.35	.65	.42	.56	.35
SN	.75	.80	.81	.87	.73	.86	.77	.86
TH	.96	.93	.95	.96	.94	.89	.95	.97
TV	.97	.97	.97	.97	.97	.97	.97	.97
LV	.93	.94	.92	.97	.90	.92	.91	.97

Table 2. ICCs for ordinates between expert and automated localization of SCG displayed for each combination of method/threshold and structure

ICCs on ordinates (No. = 10)	Threshold = 0 Methods				Threshold = 1 Methods			
	1	2	3	4	1	2	3	4
PT	.91	.91	.91	.93	.92	.89	.91	.90
CN	.92	.91	.92	.63	.92	.90	.91	.62
SN	.98	.99	.99	.99	.98	.99	.99	.99
TH	.96	.97	.98	.98	.95	.96	.97	.98
TV	.92	.93	.96	.49	.93	.95	.96	.49
LV	.99	.99	.99	.99	.99	.99	.99	.99

As mentioned in section 2.5, these results are not that easy to comprehend, especially when we try to extract trends of each parameter effect from Table 1 and 2. These tables are thus rearranged in a different configuration and then visualized so that the image generated provides a synoptic presentation of results. The right color bar reflects the range of possible ICCs value in the whole dataset. Trends on parameter effects can then be recovered regarding the warmth of a color coded cell. Interpretations regarding methods or structures will focus more particularly on color gradations respectively over lines or columns. The way this color gradation evolves over the whole image might also help comprehending possible interaction between parameters (Fig. 8).

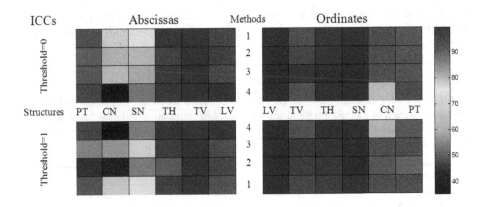

Fig. 8. Visualization of ICCs for both abscissas and ordinates of SCG regarding each parameter of the process; method (1/2/3/4), structure (PT/CN/SN/TH/TV/LV) and threshold (0/1)

Helping on the right color bar, one can easily argue that ordinates were determined more accurately than abscissas. Applying a defuzzification threshold appeared providing a very slight improvement in some results. For abscissas, caudate nucleus (CN) and subthalamic nucleus-substantia nigra (SN) appeared both outlying with lowest ICCs values. pyramidal-mamillothalamic tract (PT) was best localized using method 2 associated with a defuzzification threshold. Thalamus (TH) and lateral ventricle (LV) seemed best localized using method 4. The third ventricle (TV) was well localized regardless of the method used. For ordinates, results with method 4 on TV and CN appeared really mediocre. Localization accuracy of PT seemed insensitive to the method applied. Lastly, TH seemed best localized with the last versions of methods applied (3 and 4) while results on ordinates localization of LV and SN appeared very good regardless of the method.

3.3   Comparisons of TCs Between Structures Mapping Procedures

In a pure descriptive purpose, the average value of TC on the entire dataset that was 33.0 (given as percents) without applying a defuzzification threshold and 36.2 otherwise. Describing overlapping of both procedures on MRI, regarding methods, mean TC (and between parentheses, the same value applying a defuzzification threshold) ranged as follows: 35.9 for method 4 (38.4), 35.8 for method 3 (36.5), 31.5 for method 1 (35.8) and 29.1 for method 2 (34.2). Regarding now structures, mean TC ranged as follows: 63.5 for lateral ventricle (67.8), 36.8 for thalamus (41.5), 31.0 for third ventricle (39.7), 26.8 for both subthalamic nucleus-substantia nigra (28.1) and pyramidal-mamillothalamic tracts (25.8) and 12.6 for caudate nucleus (14.5).

Analyzing now the whole dataset and using its pure within-subject design accounting for repeated measures in the same patient, interaction between method and structure was significant ($p < 0.0001$). Adjusted to one another, effects of method ($p = 0.0056$) and structure ($p < 0.0001$) upon TC were significant, but these cannot be reported simply as main effects since their interaction is qualitative, i.e. methods, regarding their efficiency on TC, are not sorted in the same order for every structure. Applying a defuzzification threshold in the algorithm showed also an adjusted significant effect (p=0.0065). We stratified remainder analyzes on this factor too which is referred to as "threshold" effect. These comparisons of TC, first between structures - given a method- and between methods -given a structure- could be reported in two tables. This was done so, but only for the former analysis in Table 3.

Focusing on method effect, it was not significant for both CN and TV regardless of the defuzzification threshold. Summarizing results, method 4 appeared the best option regarding TC for most structures except for SN, even if differences with method 2 and 3 were not significant for that particular structure.

Focusing now on structure effect (which could have been displayed in a table similar to Table 3), cerebrospinal fluid appeared systematically the most easily labeled, expect for the third ventricle (TV) when a research of connected components is applied (method 4). TC appeared almost systematically significantly higher for LV while CN presented significantly lower TC values. Methods did not sort in the same order regarding TC values, for structures well localized, i.e. LV and TH, method 4 seemed significantly better working with a significant enhancement when a defuzzification threshold is applied, except for PT.

Threshold effect was significant in the analysis, it induced an improvement for most structure labeling except for PT due to its anatomical shape which appears particularly elongated on MRI coronal slices and SN which pixels on MRI appeared already scattered.

Viewing results (Fig. 9), TCs values computed on LV appeared outstanding compared with other structures. On the contrary, the automated procedure provided very poor results on CN. Upper and lower parts of the image appeared symmetric but with warmer colors when a defuzzification threshold is applied. The same exception is noticed about pixels "PT × method 4" and "SN × method 3" where applying a defuzzification threshold seemed slightly deleterious. For LV, methods sort in a particular order regarding TCs with in descending order method 4, 3, 2 and at last method 1. This order was not the same for all structures, revealing a qualitative interaction between method and structure effects.

Results for Tukey's honestly significant difference (THSD) test were displayed in figure 9, with vertical or horizontal strips of the image.

Two examples were represented applying for both a defuzzification and using common underscores as a symbol for insignificant differences.

Results for method effect with a particular structure (SN) showed that method 1 provided results significantly lower compared with method 2, 3 or 4 which were not significantly different.

Results for structure effect with a particular method (method 1) showed that LV provided results significantly higher compared with other structures. Furthermore, TCs was significantly higher for TV compared with CN, SN and PT which were not significantly different, TCs for TH was significantly higher compared with CN, SN and lastly, TCs were not significantly different for CN, SN and PT.

Table 3. Method effect on TC (percent) for each combination of structure/threshold factors, p-values refer to the single method effect, Tukey's honestly significant difference (THSD) test is computed when appropriate. Means with the same letter are not significantly different.

	No.	Method	Mean	THSD		Method	Mean	THSD	
			Threshold = 0				**Threshold = 1**		
			p<0.0001				p<0.0001		
	10	4	40.33	A		4	34.81	A	
PT	10	3	25.37	B		3	26.75	B	
	10	1	24.95	B		1	25.59	B	
	10	2	16.45		C	2	16.18		C
			p=0.2815				p=0.1262		
	10	2	15.92	.		2	20.53	.	
CN	10	1	15.28	.		1	17.82	.	
	10	3	12.88	.		3	13.65	.	
	10	4	06.13	.		4	05.91	.	
			p=0.0134				p=0.0005		
	10	3	40.33	A		2	32.64	A	
SN	10	4	25.37	A B		3	30.33	A	
	10	2	24.95	A B		4	28.01	A	
	10	1	16.45	B		1	21.32	B	
			p=0.0386				p=0.0009		
	10	4	45.50	A		4	56.35	A	
TH	10	1	34.67	A		1	39.41	B	
	10	3	33.43	A		3	35.12	B	
	10	2	33.27	A		2	35.05	B	
			p=0.5551				p=0.1933		
	10	3	35.59	.		2	44.83	.	
TV	10	2	33.92	.		1	43.28	.	
	10	1	32.49	.		3	42.76	.	
	10	4	25.54	.		4	27.72	.	
			p<0.0001				p=0.0001		
	10	4	72.31	A		4	77.42	A	
LV	10	3	66.98	A		3	70.10	A	
	10	1	64.26	A		1	67.04	A B	
	10	2	50.33	B		2	56.22	B	

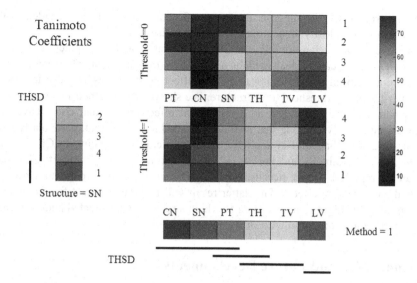

Fig. 9. Visualization of Tanimoto coefficients regarding each parameter of the process; method (1/2/3/4), structure (PT/CN/SN/TH/TV/LV) and threshold (0/1). Tukey's honestly significant difference (THSD) were displayed were displayed for 2 examples, using common underscores for insignificant differences.

4 Discussion

At this step of our procedure development, we succeeded in including indicators that are providing relevant information, qualifying and quantifying the enhancement of any option change in our automated algorithm. Thus, both ICCs and TCs provided objective indications to differentiate between methods that are developed in our procedure. Methods are sorted in an order that could not be predicted neither prior to assessment nor after running the process since trying crude comparisons in such a topic highlights its complexity. It is for example a bit striking that method 3, which account for directional relationships in a quite simple way, did not provide the worse results. Actually, the feature of results makes them a bit difficult to interpret, because even if ICCs provide an order in methods accuracy as compared with expert procedure, TCs explore another capability of the automated procedure which may change this order. Furthermore, results are not unequivocal since the improvement of a particular method compared to another is not stable depending on which structure is treated.

ICCs showed that while localization accuracy of structures is on the whole acceptable, abscissas seemed a bit more scattered around identity line with lower values as compared with ordinates. This could be explained by the structures overlapping in the direction of abscissa axis, with a smaller range of SCG as compared with ordinates. ICCs also showed that some structures like CN appeared more difficult to localize, may be due to the pixel dispersion obtained with our procedure (Fig. 5) and also to its small size that may constitute a weak point in our process which tend to homogenize sizes of structures. This suggests that a specific correction will be provided to this structure which is clearly presenting a systematic offset (Fig. 7).

TC appeared a much more relevant criterion in that it can be computed for each observation and that can be dealt with as a ratio level measurement. This enabled us to carry out a factorial design to assess method effect on its values and also to account for other factors as threshold and structure. We showed that method factor, threshold factor and structure factor had all a significant effect on TC that could have been summarized roughly noticing that method 4 seemed more efficient, particularly when a defuzzification threshold is applied and that structures belonging to CF are more easily outlined (Table 3 and Fig. 9). What TC taught us more is that these effects cannot be reported simply because of a significant qualitative interaction between method and structure effects. This latter result will make us account for this interaction splitting the algorithm according to which structure is segmented and/or enhancing expert prior information with such new constraints.

5 Conclusion and Future Developments

Our automated procedure is already promisingly working. Including statistical criterions (e.g. TC) in that phase of selecting option will enhance objectively our process permitting statistical inference on whether changing option in the procedure leads to a significant enhancement or not. Furthermore, our research is challenging to use these criterions not only in that "calibration" phase of our process development, but to include it similarly as a fit statistic associated with a process of connected component research and morphologic features. Therefore, we have now relevant indicators that will help us design and assess our automated segmentation procedure in further developments. Using visualization methods for such complex results will be particularly useful for a synoptic and quick comprehension of factor effects emphasizing general trends. This method will help us exploring further results; including defuzzification threshold as continuous variable or using specific processes regarding structure adding modalities for the method factor in structure where results keep insufficient.

References

1. Coubes P., Vayssiere N., El Fertit H., Hemm S., Cif L., Kienlen J., Bonafe A., Frerebeau P.: Deep Brain Stimulation for Dystonia: Surgical Technique. Stereotact Funct Neurosurg. 78 (2002) 183-191
2. Walter B.L., Vitek J.L.: Surgical treatment for Parkinson's disease. Lancet Neurol. 3 (2004) 719-728
3. Caire F., Derost P., Coste J., Lemaire J.-J.: Stimulation Sous-Thalamique dans la Maladie de Parkinson Sévère: Étude de la Localisation des Contacts Effectifs. Neurochirurgie. (2005) in press
4. Lemaire J.-J., Durif F., Boire J.-Y., Debilly B., Irthum B., Chazal J.: Direct Stereotactic MRI Location in the Globus Pallidus for Chronic Stimulation in Parkinson's Disease. Acta Neurochir (Wien). 141 (1999) 759-766

5. Bejjani B.P., Dormont D., Pidoux B., Yelnik J., Damier P., Arnulf I., Bonnet AM., Marsault C., Agid Y., Philippon J., Cornu P.: Bilateral Subthalamic Stimulation for Parkinson's Disease by Using Three-Dimensional Stereotactic Magnetic Resonance Imaging and Electrophysiological Guidance. J Neurosurg; 92 (2000) 615-25

6. Villéger A., Lemaire J.-J., Boire J.-Y.: Localization of Target Structures through Data Fusion Applied to Neurostimulation. Proc of the IEEE Engineering in Medicine and Biology Society 27th Annual International Conference, Shanghai) (2005)

7. Villéger A.: Fusion de Données et Raisonnement Spatial en Vue de la Localisation de Structures Profondes en Imagerie Cérébrale. Ph D thesis. Auvergne University (2005)

8. Barra V., Boire J.Y.: Tissue Segmentation on MR Images of the Brain by Possibilistic Clustering on a 3D Wavelet Representation. Journal of Magnetic Resonance Imaging, 11 (2000) 267-278

9. Lévy PP.: The case view, a generic method of visualization of the case mix. International Journal of Medical Informatics, 73, (2004) 713-718

10. Lévy PP., Duché L., Darago L., Dorléans Y., Toubiana L., Vibert JF., Flahault A.: ICPCview: visualizing the International Classification of Primary Care. Studies in Health Technology and Informatics. 116 (2005) 623-628

11. Smith C.A.B.: On the Estimation of Intraclass Correlation. Annals of Human Genetics, 21, (1956) 363-373

12. Bartko J. J.: The Intraclass Correlation Coefficient as a Measure of Reliability. Psychological Reports. 19 (1966) 3-11

13. Shoukri M.M., Pause C.A.: Statistical Methods for Health Sciences. 2nd edn. CRC Press, Boca Raton (2000)

14. Tanimoto T.T.: IBM Internal Report. (1957)

15. Keppel G.: Design and Analysis: A Researcher's Handbook. 3rd edn. Prentice Hall, New Jersey (1991)

16. SAS, Statistical Analysis System, v8.02, SAS Institute Incorporation, Cary, NC, USA

17. Fitzmaurice G.M., Laird N.M., Ware J.H.: Applied Longitudinal Analysis. Wiley, New Jersey (2004)

18. Tukey J.W.: The problem of Multiple Comparisons, unpublished manuscript (1953). In SAS Institute Inc. SAS/STAT User's Guide, Version 8, 1st edn. Vol. 3. Cary, NC, USA (2002)

19. MATLAB, v7.01 (2004), The MathWorks Incorporation

A Computational Method for Viewing Molecular Interactions in Docking

Vipin K. Tripathi, Bhaskar Dasgupta, and Kalyanmoy Deb

Indian Institute of Technology, Kanpur U.P. 208016, India
{vipinkt,dasgupta,deb}@iitk.ac.in

Abstract. A huge amount of molecular data is available in protein data bank and various other libraries and this amount is increasing day by day. Devising new and efficient computational methods to extract useful information from this data is a big challenge for the researchers working in the field. Computational molecular docking refers to computational methods which attempt to obtain the best binding conformation of two interacting molecules. Information of the best binding conformation is useful in many applications such as rational drug design, recognition, cellular pathways, macromolecular assemblies, protein folding etc. Docking has three important aspects: (i) modeling of molecular shape, (ii) shape matching and (iii) scoring and ranking of potential solutions. In this paper, a new approach is proposed for shape matching in rigid body docking. The method gives visual information about the matching conformations of the molecules. In the approach proposed here, B-spline surface representation technique is used to model the patches of molecular surface. Surface normal and curvature properties are used to match these patches with each other. The 2-D approach used here for generation of surface patches is useful to pixellisation paradigm.

1   Introduction

Molecular interactions are primitive to all the biological phenomena occurring in nature. Knowledge of molecular associations helps in understanding a variety of pathways taking place in the living cell. Molecular docking has industrial application also. It is useful to pharmaceutical industry for rational drug design. Docking of two molecules is shown in Fig. 1.

In computational molecular docking, the goal is to find the transformation (translation and rotation) which brings one molecule into optimal fit with the other molecule. The methods used for docking are generally classified into two classes, rigid body docking and flexible ligand docking. The rigid body docking methods consider both the molecules as rigid bodies whereas the flexible ligand methods consider one or both the molecules flexible.

An association of two molecules can be rated as correct or incorrect based on different criteria like shape complementarity between molecules, binding energy

P.P. Lévy et al. (Eds.): VIEW 2006, LNCS 4370, pp. 152–163, 2007.

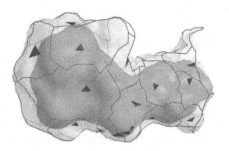

Fig. 1. Docking of two molecules

of a complex, electrostatic and van der Waals interactions etc. Although other criteria are also important, shape complementary is considered a necessary condition in rigid body docking. Shape complementarity criterion assumes that the molecular surfaces of the two molecules need to match if the molecules are to bind to each other with high affinity. This approach was initially adopted by Kuntz and his colleagues [10] and was further elaborated and rationalized by Connolly [4].

Docking is computationally difficult because there are many alternate binding locations and orientations in which two molecules can bind with each other. The number of possibilities grow exponentially with the size of the molecules. Large size molecules, non-exact nature of the structure of molecules and conformational changes taking place upon association add more complexity to the problem. These difficulties make docking problem one of the most challenging problems in structural bioinformatics.

Complexities involved in a docking problem create the need for a simple, yet, efficient docking procedure which can provide some preliminary solutions for detailed further analysis. Visualization of docking conformations is an extra help provided by a method. A simple shape representation technique with an efficient shape matching strategy are the basic requirements for any such docking procedure. B-spline surface approximation technique offers a lot of flexibility required for fitting a surface through scattered data [14]. It is, therefore, a promising shape representation technique for molecular docking. In this paper, we propose a simple and efficient shape matching scheme for rigid body docking which uses B-spline surface fitting technique for shape representation. Molecular surface is considered as a network of connected subregions known as 'patches'. These patches on the surface of outer molecule are indicated by small triangles in Fig. 1. The lines indicate the boundaries of the patches. Use of B-spline representation for modeling the molecular surface patches is a new approach adopted here. Shape complementarity is considered here as criterion for docking. For shape matching, patches of the molecular surfaces with complementary curvatures are aligned with each other. Normals at the surface points play an important role in the matching strategy.

2   Past Studies

Traditionally, the two major approaches have been adopted in a docking problem. The first is a geometry based approach while the second focuses on the energy of binding. The geometry based approach was initially adopted by Kuntz and his colleagues [10] and was further elaborated and rationalized by Connolly [4].

Various shape representation schemes have been used by different researchers. A complete review of computation and visualization of surfaces is given in [5]. Connolly's analytical surface represents the molecular surface as a surface connected by a network of convex, concave and saddle shape surfaces which smoothes over the crevices and pits between the atoms [3]. A description of surface by sparse critical points is computed by some of the researchers [12,13,16]. Sanner et al. [19] compute a reduced surface in their MSMS algorithm and from reduced surface solvent-accessible and solvent-excluded surfaces are computed. Some 'shape-implicit' methods have also been used which discretize the molecule onto a grid in space and consider only the occupied cells, i.e., voxels, to define the shape of the molecule [17,9]. B-spline surface fitting has been used earlier for fitting surface through scattered data. For related studies [6,20] can be referred. Using B-spline surface fitting for molecular surface representation is a new approach adopted here.

Docking algorithms may be classified into two broad classes: (i) brute force enumeration algorithms, (ii) local shape feature matching algorithms. Local shape feature matching algorithms have been pioneered by Kuntz [10]. These algorithms use complementary features of surfaces to align them. They include algorithms like, distance geometry algorithms [10,8], pose clustering [15,9], geometric hashing [21,7] and geometric hashing and pose clustering [18,11].

3   Shape Representation

A geometric surface consists of a network of surface patches of different forms such as convex, concave , flat, etc. connected to each other. Use of B-spline surface representation technique is proposed in this paper to generate the patches of molecular surface. Those patches are used in a patch by patch shape matching procedure. The B-spline surface is controlled by a characteristic polyhedron as shown in Fig. 2. Advantages of this surface are the local control of surface and independence of the degree of surface from number of control points. The surface generation scheme requires molecular data already parameterized on a 2-D grid as input. The rectangular data grid is shown in Fig. 3. An inverse problem in B-spline surface fitting is solved in the process of generating a surface from 2-D surface point data.

A tensor product B-spline surface is formulated as

$$\mathbf{Q}(u,v) = \sum_{i=0}^{n} \sum_{j=0}^{m} \mathbf{P}_{ij} N_{i,k}(u) N_{j,l}(v) \qquad (1)$$

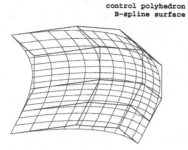

Fig. 2. B-spline surface   **Fig. 3.** Rectangular 2-D grid of data points

where,

$$\mathbf{Q}(u, v) = \text{a point on the surface,}$$
$$\mathbf{P}_{ij} = \text{vertices of the defining polyhedron,}$$
$$N_{i,k}(u),\ N_{j,l}(v) = \text{Blending functions in u and v directions, respectively.}$$

The B-spline blending functions are computed by the following recursive expressions:

$$N_{i,1}(u) = \begin{cases} 1 \text{ if } t_i \leq u < t_{i+1}, \\ 0 \text{ otherwise.} \end{cases} \tag{2}$$

$$N_{i,k}(u) = \frac{(u - t_i)N_{i,k-1}(u)}{t_{i+k-1} - t_i} + \frac{(t_{i+k} - u)N_{i+1,k-1}(u)}{t_{i+k} - t_{i+1}} \tag{3}$$

The t_i are known as knot values. They relate the parametric variable u to the control points $\mathbf{P}_{ij}$. For an open curve t_i are given by

$$t_i = \begin{cases} 0 & \text{if } i < k, \\ i - k + 1 & \text{if } k \leq i \leq n, \\ n - k + 2 & \text{if } i > n, \end{cases} \tag{4}$$

with

$$0 \leq i \leq n + k. \tag{5}$$

The range of parametric variable u is

$$0 \leq u \leq n - k + 2. \tag{6}$$

Finding $\mathbf{Q}$, the surface data points, by an input of $(n + 1) \times (m + 1)$ number of control points which are the vertices of control polyhedron $\mathbf{P}$, is known as the 'forward problem'.

For shape representation the inverse problem is of interest, i.e., given a known set of data points on a surface, determine the defining polyhedron for B-spline surface that best approximates the data. The control points obtained after solving the inverse problem, being very less in number as compared to the given data points, are used for generation and processing of surface patches. This needs less storage and processing time. Major steps of the method used here for solving the inverse problem are given below.

Writing out Eq. 1 for a single surface data point, say, point $\mathbf{Q}(u_\alpha, v_\beta)$ as shown in Fig. 3 yields

$$\mathbf{Q}(u_\alpha, v_\beta) = N_{0,k}(u_\alpha)[N_{0,l}(v_\beta)\mathbf{P}_{00} + N_{1,l}(v_\beta)\mathbf{P}_{01} + \cdots + N_{m,l}(v_\beta)\mathbf{P}_{0m}] +$$

$$\vdots$$

$$N_{n,k}(u_\alpha)[N_{0,l}(v_\beta)\mathbf{P}_{n0} + N_{1,l}(v_\beta)\mathbf{P}_{n1} + \cdots + N_{m,l}(v_\beta)\mathbf{P}_{nm}]. \quad (7)$$

For (div_u+1)*(div_v+1) data points, above equation can be written in matrix form as,

$$[Q]_{(div\_u+1),(div\_v+1)} = [C^t]_{(div\_u+1),(n+1)} * [P_{ij}]_{(n+1),(m+1)} * [D]_{(m+1),(div\_v+1)}. \quad (8)$$

This is simplified to,

$$q = [A]p \quad (9)$$

where,
q =Column vector of data points of dimension
     (div_u+1)*(div_v+1),1,
p =Column vector of control points of dimension
     (n+1)*(m+1),1,
[A] =Matrix of dimension
     (div_u+1)*(div_v+1),(n+1)*(m+1).

Matrix [A] is formed as,

$$A_{I,J} = d_{J,I}[C^t] \quad (10)$$

where, $d_{(J,I)} = (J,I)^{th}$ element of D matrix.

As the surface point data available in molecular docking problem will give a highly over-specified matrix $\mathbf{A}$, system $[\mathbf{A}^T\mathbf{A}]\mathbf{p} = [\mathbf{A}^T]\mathbf{q}$ is solved instead of the system $\mathbf{A}\mathbf{p} = \mathbf{q}$, using Cholesky decomposition of matrix $\mathbf{A}$. Matrix $[\mathbf{A}^T\mathbf{A}]$ on the left hand side is a square matrix then. If Cholesky decomposition fails due to matrix $[\mathbf{A}^T\mathbf{A}]$ not being positive definite, a method based on singular value decomposition is used which is applicable to a general rectangular $M \times N$ matrix. For detailed discussion related to procedure used for the solution of inverse problem [20] can be referred.

4   Shape Matching

After solving the inverse problem, the control points of the two surface patches one from each of the two molecules are available. These control points will now be used for surface generation and other processing. Hence, the problem of surface matching can now be described in terms of the control points of the surface patches.

4.1 Description of Problem

Given the control points of two surface patches, check whether the patches are complementary. If the patches are complementary, Compute:

1. The translation required to match two given surface patches at a desired point.
2. The rotation required to match the surface patches in the same orientation.

5 Solution Strategy and Formulation

Shape matching of molecular surfaces is done by a patch by patch matching strategy. The molecular surface consists of patches of different types, like convex, concave, flat, etc. But, out of these patches, only those which are complementary in nature, like a pair of concave and convex patches, will demonstrate an effective match from molecular docking point of view. Therefore, the first requirement is to check whether the surface patches are complementary or not. The algorithm proposed here matches the complementary surface patches at a desired point and in desired orientation.

5.1 Check for Complementary Surface Patches

To check if the surface patches are complementary, the principal curvatures of both the surface patches are calculated by Eq. 11.

$$\kappa_{max} = H + \sqrt{H^2 - K}$$
$$\kappa_{min} = H - \sqrt{H^2 - K}. \tag{11}$$

In Eq. 11, H is mean curvature and K is Gaussian curvature [14].

If

κ_{max1} = maximum curvature of the first surface patch at the desired point,

κ_{min1} = minimum curvature the first surface patch at the desired point,

κ_{max2} = maximum curvature of the second surface patch at the desired point,

κ_{min2} = minimum curvature the second surface patch at the desired point,

the surface patches are **not** complementary when $\kappa_{max1} \times \kappa_{max2} > 0$ or $\kappa_{min1} \times \kappa_{min2} > 0$.

5.2 Obtaining Translation and Rotation for Complementary Surfaces

Translation. For matching the surface patches at a given point on both the surface patches, the strategy used here is to translate both the surface patches to the origin of the coordinate frame such that the desired points on both the surface patches coincide with the origin.

The translation for the first surface patch is computed by the following equation

$$\mathbf{P}_{1ijT} = \mathbf{P}_{1ij} - \mathbf{Q}_1 \tag{12}$$

where,

$\mathbf{P}_{1ij}$ = control points of the first surface patch,
$-\mathbf{Q}_1$ = translation for the first surface patch,
$\mathbf{P}_{1ijT}$ = changed control points of the first surface patch after translation .

New control points will generate the surface patch in translated position, i.e., at the origin. This is due to the property of B-spline basis functions known as 'partition of unity'.

Translation for the second surface Patch is computed by

$$\mathbf{P}_{2ijT} = \mathbf{P}_{2ij} - \mathbf{Q}_2 \tag{13}$$

where,

$\mathbf{P}_{2ij}$ = control points of the second surface patch,
$-\mathbf{Q}_2$ = translation for the second surface patch,
$\mathbf{P}_{2ijT}$ = changed control points of the second surface patch after translation .

Changed control points will generate the second surface patch in translated position such that the desired point on the second surface patch will coincide with the origin and thus, with the desired point on the first surface patch. This completes the translation of surface patches.

Rotation. Translation brings the surface patches at the origin where they meet each other at a point. But it does not change their orientation. The next task is to find the required rotation which will reorient and superimpose one surface patch over the other. This is required to judge as to how well they match with each other.

The strategy used to match the surface patches in the same orientation is, to compute a triad of unit vectors at the desired point on both the surface patches and then to match these unit vectors of a triad with the coordinate axes. For a convex surface patch, normal $\mathbf{n}$ is made to align with negative z axis while for a concave surface patch it coincides with positive z axis. Noting that, after translation desired points are already coincident with the origin of the coordinate frame, this brings the surface patches in the same orientation.

A triad at the desired point on any of the two surface patches consists of three unit vectors: a unit tangent vector in u parametric direction $\mathbf{q}_u$, a unit normal to the surface patch $\mathbf{n}$ and a unit vector $\mathbf{m}$ normal to both of $\mathbf{q}_u$ and $\mathbf{n}$ (refer [14]). A matrix $\mathbf{V}$ is formed by writing vectors of a triad in the rows of the matrix and a matrix $\mathbf{W}$ is formed similarly by the coordinate axes vectors. Since $\mathbf{V}$ and $\mathbf{W}$ are unit vector matrices, it can be written

$$[\mathbf{x}\ \mathbf{y}\ \mathbf{z}]^T = [\mathbf{q}_{u1}\ \mathbf{m}_1\ \mathbf{n}_1]^T[\mathbf{R}].$$

Then, required rotation matrix for system $\mathbf{V}$ with respect to system $\mathbf{W}$ is simply

$$[\mathbf{R}] = [\mathbf{V}]^{-1}[\mathbf{W}]. \tag{14}$$

Using Eq. 14 now, changed control points of the surface patches after rotation can be obtained. For the first surface patch,

$$\mathbf{P}_{1ijR} = \mathbf{P}_{1ijT} \times \mathbf{R}_1 \tag{15}$$

where,

$$\mathbf{R}_1 \quad = \text{rotation matrix for the first surface patch}$$
$$\mathbf{P}_{1ijR} = \text{changed control points of the first surface patch after rotation .}$$

Similarly, for the second surface patch,

$$\mathbf{P}_{2ijR} = \mathbf{P}_{2ijT} \times \mathbf{R}_2 \tag{16}$$

where,

$$\mathbf{R}_2 \quad = \text{rotation matrix for the second surface patch}$$
$$\mathbf{P}_{2ijR} = \text{changed control points of the second surface patch after rotation.}$$

Changed control points will generate the surface patches in rotated position. If both the surface patches are generated with the changed control points and plotted on the same plot, they match in the same orientation.

6   Proposed Algorithm

The algorithm proposed in this section for matching two surface patches at a desired point and in desired orientation is as follows:

step 1: Compute the principal curvatures of both the surface patches and check for complementarity as in section 5.1.

step 2: If surface patches are complementary, input the values of parameters u and v and the control points $\mathbf{P}_{1ij}$ for the first surface patch. Find translation $\mathbf{Q}_1$ by Eq. 1.

step 3: Obtain changed control points of the first surface patch after translation $\mathbf{P}_{1ijT}$, using Eq. 12.

step 4: Compute triad $(\mathbf{q}_{u1}, \mathbf{m}_1, \mathbf{n}_1)$ and form matrix $\mathbf{V}$ as $[\mathbf{q}_{u1}, \mathbf{m}_1, \mathbf{n}_1]^T$.

step 5: If principal curvatures of the first surface patch are positive, form matrix $\mathbf{W}$ as $[\mathbf{x}, \mathbf{y}, \mathbf{z}]^T$ and if they are negative, form matrix $\mathbf{W}$ as $[\mathbf{x}, \mathbf{y}, -\mathbf{z}]^T$.

step 6: Find rotation matrix $\mathbf{R}_1$, using Eq. 14.

step 7: Obtain changed control points of the first surface patch after rotation $\mathbf{P}_{1ijR}$, using Eq. 15.

step 8: Repeat steps 3 to 7 for the second surface patch.

step 9: Generate both the surface patches using changed control points after rotation $\mathbf{P}_{1ijR}$ and $\mathbf{P}_{2ijR}$ in Eq. 1 one by one. Plot both the surface patches in the same plot to see the match.

7   Simulation Results and Discussion

A few results of testing the algorithm given in the previous section on different types of surface patches are shown here. The surface patches are generated using method given in section 3. The input surface point data sets are already available as parameterized data on a 2-D rectangular grid. The algorithm is, yet, to be tested on actual protein data available in protein data bank. For generation of surfaces, 651 data points have been used in a data set. The curves in u and v parametric directions are cubic. The algorithms are coded in C programming language and the results are displayed in *gnuplot* in *linux* environment.

Fig. 4 to Fig. 6 represent the surface patches from the first molecule. Fig. 7 to Fig. 9 represent the surface patches from the second molecule. Fig. 10 to Fig. 12 show the matches of two surface patches selected from two different molecules. No range of data is fixed here as *gnuplot* automatically sets relative sizes in this case. The view angle is chosen such that the match is visible clearly. It can, easily, be seen that Fig. 10 shows a better binding conformation of two surface patches than that shown in Fig. 11 and Fig. 12.

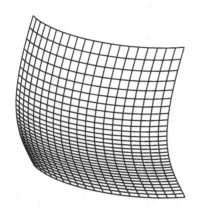

Fig. 4. Surface patch no. 1 of the first molecule

Fig. 5. Surface patch no. 2 of the first molecule

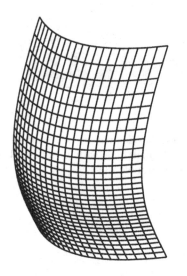

Fig. 6. Surface patch no. 3 of the first molecule

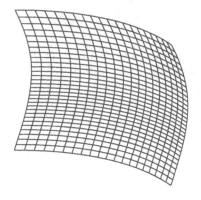

Fig. 7. Surface patch no. 1 of the second molecule

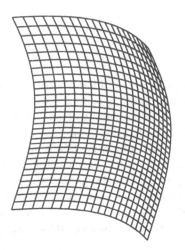

Fig. 8. Surface patch no. 2 of the second molecule

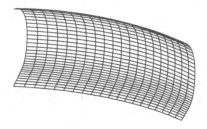

Fig. 9. Surface patch no. 3 of the second molecule

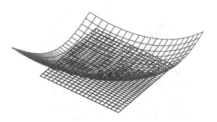

Fig. 10. Surface patch no. 1 of the first molecule matching with patch no. 1 of the second molecule

Fig. 11. Surface patch no. 2 of the first molecule matching with patch no. 1 of the second molecule

Fig. 12. Surface patch no. 1 of the first molecule matching with patch no. 3 of the second molecule

8   Conclusion

The B-spline surface representation effectively represents different types of surface patches of molecular surface. The shape matching algorithm provides important visual information by matching the surface patches at desired location and in desired orientation. It is easy to compare the binding conformations simply by viewing the results. Shape representation and computing triads for shape matching is simple. Matching strategy uses single point from each of the surface patch so problem of non existence of matching point pairs between the receptor and the ligand does not arise which is the case with some strategies used earlier in the literature. Shape matching based on the matching of triads reduces the number of potential candidate transformations. So, this gives an efficient shape matching algorithm with less complexity.

References

1. Ausiello, G., Cesareni G., Helmer-Citterich: M ESCHER: A new docking procedure applied to the reconstruction of protein tertiary structure. Proteins **28** (1997) 556–567
2. Chen, R., Weng, Z.: Docking unbound proteins using shape complementarity, desolvation and electrostatics. Proteins **47** (2002) 281–294

3. Connolly, M.: Analytical molecular surface calculation. J. of Appl. Cryst. **16** (1983) 548–558
4. Connolly, M.: Shape complementarity at the hemoglobin $\alpha_1\beta_1$ subunit interface. Biopolymers **25** (1986) 1229–1247
5. Connolly, M.: Molecular surfaces: a review. www.chem.leeds.ac.uk/ICAMS (1996)
6. Eck, M., Hoppe, H.: Automatic reconstruction of B-spline surfaces of arbitrary topological type. SIGGRAPH '96: Proc. 23rd anal. conf. on comp. graph. and int. tech., New York (1996) 325–334
7. Fischer, D., Lin, S. L., Wolfson, H. J., Nussinov, R.: A geometry-based suite of molecular docking processes. J. of Mol. Biol. **248**(1995) 459–477
8. Gardiner, E. J., Willet, P., Artimiuk, P. J.: Protein docking using a genetic algorithm. Proteins **44**(2001) 44–56
9. Goldman, B. B., Wipke, W. T.: QSD quadratic shape descriptors. 2. Molecular docking using quadratic shape descriptors (QSDock). Proteins **38**(2000) 79–94
10. Kuntz, I., Blaney, J., Oatley, S., Langridge R., Ferrin T.: A geometric approach to macromolecule-ligand interactions. J. of Mol. Biol. **161** (1982) 269–288
11. Lenhof, H.: Parallel protein puzzle: a new suite of protein docking tools. Proc of the first ann int conf on comp mol biol RECOMB 97 (1997) 182–191
12. Lin, S. L., Nussinov, R., Fischer, D., Wolfson, H. J.: Molecular surface representation by sparse critical points. Proteins **18** (1994) 94–101
13. Lin, S. L., Nussinov, R.: Molecular recognition by face center representation of a molecular surface. J. of Mol. Graph. **14**(1996) 78–90
14. Mortenson, M. E.: Geometric Modeling. John Wiley and Sons (1985)
15. Norel, R., Lin, S. L., Wolfson, H., Nussinov, R.: Shape complementarity at protein-protein interfaces. Biopolymers **34** (1994) 933–940
16. Norel, R., Petrey, D., Wolfson, H., Nussinov, R.: Examination of shape complementarity in docking of unbound proteins. Proteins **35** (1999) 403–419
17. Palma, P. N., Krippahl, L., Wampler, J. E., Moura, J.G.: BIGGER: a new (soft) docking algorithm for predicting protein interactions. Proteins **39** (2000) 372–384
18. Sandak, B., Nussinov, R., Wolfson, H.: A method for biomolecular structural recognition and docking allowing conformational flexibility. J. of Comp. Biol. **5** (1998) 631–654
19. Sanner, M. F., Olson, A. J., Spehner, J. C.: Reduced surface: an efficient way to compute molecular surfaces. Biopolymers **38** (1996) 305–320
20. Tripathi, V. K., Dasgupta, B., Deb, K.: An approach for surface generation from measured data points using B-splines. Int. Conf. on Systemics, Cybernetics and Informatics, Hyderabad, India (2005) 578–581
21. Wolfson, H.: Model based object recognition by 'Geometric Hashing'. Proc. of the Euro. Conf. on Comp. Vision **25** (1990) 526–536

A Graphical Tool for Monitoring the Usage of Modules in Course Management Systems

Riccardo Mazza

Institute for Communication Technologies
Faculty of Communication Sciences
University of Lugano – Switzerland
mazzar@lu.unisi.ch
Department of Innovative Technologies
University of applied sciences
of Southern Switzerland.
Manno - Switzerland
riccardo.mazza@supsi.ch

Abstract. This paper proposes an approach to graphically represent the tracking data in Course Management Systems in order to mine and discover the usage of specific software modules. We have implemented a tool that aims to give, at a glance, a visual representation of the usage of a specific module in all the courses managed by the Course Management System. The tool also allows comparing the usage in different courses, and seeing the distribution of the usage over time. This information is useful to administrators of the Course Management System that have to know how much the modules have been used in courses. But also, it could be useful to instructional designers that have to design, plan, and evaluate the learning needs in institutions.

1 Introduction

Currently, many educational institutions and enterprises have set up Course Management Systems (CMS), and they are organizing their courses and activities around these new technologies. CMS are a type of software application that enables instructors to distribute information and materials to students, prepare assignments and tests, engage in discussions and manage distance classes over the Internet [4]. The great majority of available CMS are represented by commercial products (WebCT and BlackBoard are the most popular). Also, free and open source solutions such as Moodle have reached a reasonable level of maturity. Moodle provides a framework for organizing courses by composing a set of tools (called "modules") in the order that students will be using them. Examples of modules are: *Assignment* (create a task with a due date and a maximum grade), *Forum* (a discussion board), *Resource* (a piece of content for the course), *Quiz* (design and build quizzes), etc.

The administrator of such a software platform faces difficulties in monitoring the module's usage in courses. In particular, if he wants to understand to what extent a particular module (for instance, *Forum*) has been used in all the courses managed by

P.P. Lévy et al. (Eds.): VIEW 2006, LNCS 4370, pp. 164–172, 2007.

the platform, he has to enter each specific course and try to make sense of the usage of the tool. Needless to say, a platform can support thousands of courses.

Most learning environments accumulate large data logs of the students' activities (tracking data), and usually provide some monitoring features that enable administrator to view some aspects of the data. For instance, the Moodle platform allows the administrators to see a textual log with any details about users, their actions, the time, and their IP addresses. However, this tracking data is complex and organized in a tabular format, which in most cases is inappropriate for the administrator's monitoring needs.

In this work we propose an approach to graphically represent the student tracking data in order to mine and discover the usage of specific modules in Moodle. We have implemented a tool that aim to give, at a glance, a visual representation of the usage of a specific module in all the courses. The tool also allows comparing the usage in different courses, and seeing the distribution of the usage during the time. This information is useful to administrators of the Moodle platform that have to know how much the modules have been used in courses. But also, it is useful to instructional designers that have to design, plan, and evaluate the learning needs in institutions.

The paper is organized as follow. Next section presents a description of the tool. Then, we look into the graphical representations generated and formulate some interesting insights that can be derived. Finally, we summarize the work completed and outline some directions for future works.

2 System Overview

The basic idea is to use log data to compute to what extent a particular Moodle's module has been used in a particular range of time. A hypothetical user of this tool might be interested in knowing the distribution of the usage of the module in a range of time. Thus, our goal is try to represent in a compact, single screen, graphical representation the distribution of the usage of a module for all courses managed by the platform. That involves visualizing a large amount of multidimensional data on current displays. Among the several techniques available in Information Visualization [1,5] for representing multidimensional data, we have chosen the Pixel-Oriented Techniques. The basic idea underlying these techniques is "*to represent as much data as possible on the screen at the same time by mapping each data value to a colored pixel of the screen and present the data values belonging to one dimension (attribute) in a separate subwindow*" [2]. We can follow this approach to map usage data to colored pixels, and courses to subwindows. The advantages of this approach are the simplicity (both in term of technical implementation and minimal cognitive workload) and the possibility to visualize large amount of data. Successful techniques have already explored pixel-oriented visualizations with more that 1 million data values [3].

A number of issues have to be considered when designing a pixel-oriented visualization technique [2]:

1. **Shape of subwindows:** subwindows are usually displayed as rectangles. Is a rectangular shape appropriate, or does an alternative exist?

2. **Visual mapping and pixel arrangement:** how to map data into visual structures? How the visual structures are arranged inside the subwindows?
3. **Color mapping:** how to map data values to color?
4. **Ordering of subwindows:** how to order the subwindows?

In then next sections we will explore each of these steps.

2.1 Shape of Subwindows

A common partitioning of the screen, that also allows good screen usage, is the rectangular shape of subwindows. Other alternative shapes have been proposed in order to optimize the distance between the pixels belonging to the dimension of one data object [2]. However, the nature of our data and the tasks doesn't require the user to analyze correlations and dependencies between different subwindows. Hence, the rectangular shapes were selected for this project.

2.2 Visual Mapping and Pixel Arrangement

Source data comes in the form of data structures, which are the results from the queries performed on the Moodle's relational database. Moodle uses a relational database (MySQL or Postgres) to store data on courses and users. In particular, a table (mdl_log) is dedicated to store all the actions performed by the users while interacting with the course tools. For each action performed by a user, Moodle registers into this table, among other things, the following fields, that will be consider in our computations:

- Time: the timestamp when a particular action was performed
- Course: the ID of the course involved in the action
- Module: the ID of the module involved in the action

This data represents the history of the actions performed by the users on the Moodle platform, and must be processed in order to derive the data structures that will be mapped into the visual structures. Data structures can be represented in the following format:

$$r_i = (ts_i, c_i, m_i) \ . \tag{1}$$

Where ts_i represents the timestamp, c_i the course ID, and m_i the module ID.

The first computation our program does is to calculate how many pixels can be contained in each subwindow. We indicate with p_{max} this number. This depends on a) the resolution of the display, b) the number of subwindows (that is the number of courses to consider). We also have to take into account some screen space dedicated to contain GUI components, and some pixels that inevitably have to be used for borders and to leave some blank space between subwindows.

We want to provide the user with the possibility to choose the granularity of the time he wants to consider in his analysis. In fact, depending on the task characteristics and range of time, it might not make sense to analyze module usage for intervals of seconds or minutes. For this reason, the user has the ability to select the intervals of discretization of time, which can be in the form of:

- 1 second,
- 1 minute,
- 1 hour,
- 1 day,
- 1 week,
- 1 month.

Each option listed above will be offered to the user only if it fits in the subwindow (for instance, seconds are considered only if the number of seconds contained in the chosen time range is $\leq p_{max}$).

The program performs some computations in order to split the range of time selected by the user in a number n of discrete intervals delimited by $\{t_0, ..., t_n\}$, where each interval $[t_i, t_{i+1})$ reflects to the discretization granularity selected by the user. The key point here is to find a good balance between the number of intervals and pixels' usage in subwindows. Because in some cases the number of intervals (n) can be quite small with respect to the number of available pixels (p_{max}), we decided to map data into blocks of pixels, instead of a single pixel. In this way, the full displayable area of the subwindow will be used.

A block of pixels represents the basic visual structure where we map the data. It is a square composed by k^2 pixels, where k is computed with the following formula:

$$k = \left\lfloor \sqrt{\frac{p_{max}}{n}} \right\rfloor.$$

(2)

Hence, each timeframe $[t_i, t_{i+1})$ is visually represented by a block of pixels that will be colored using an appropriate color mapping described later.

Another important question is how to arrange blocks within each of the subwindows. The dataset that we are considering has a natural order that is the sequence of time frames ordered over the time. This ordering is important to the user, as he probably wants to analyze usage data over time. Because of this, we decided to use the spiral arrangement that has already been explored and has given good results in other

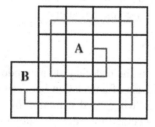

Fig. 1. The spiral placement model for the location of blocks of pixels. Block "A" describe usage data for the most recent timeframe. Block "B" describes usage data for last recent timeframe.

works [3]. Spiral placement consists to order the blocks in a rectangular spiral shape, where the center of the spiral corresponds to the first timeframe $[t_0, t_1)$, moving towards the outside of the subwindow (see Fig. 1).

2.3 Color Mapping

The color of the block is the graphical property that we use to map the level of usage of a particular tool on a given timeframe. To give the extent of the level of usage of a module on a given timeframe, we compute the number of the accesses made by the users to the module m_h during the timeframe $[t_i, t_{i+1})$ for the course c_k. That is:

$$U_{t_i,m_h,c_k} = \underset{j}{count}\{(ts_j,c_j,m_j) \mid t_i \le ts_j < t_{i+1} \wedge c_j = c_k \wedge m_j = m_h\} . \tag{3}$$

The distribution of values for U is then mapped to a scale of colors. We decided to use the HSB (Hue, Saturation, Brightness) color model with a monotonic increasing/decreasing of the saturation, full brightness, and a hue that can be chosen by the user among a fixed set. Hence, the usage of a module in a particular timeframe is mapped with the color saturation (max usage = max saturation, no usage = no saturation).

For the color saturation we provided two different functions that can be selected by the user depending on which task he has to perform. A linear mapping, in this form:

$$SAT_{t_i,m_h,c_k} = \frac{U_{t_i,m_h,c_k}}{\underset{i,h,k}{\max}(U_{t_i,m_h,c_k})} . \tag{4}$$

And a logarithmic mapping, in this other form:

$$SAT_{t_i,m_h,c_k} = \frac{\log(U_{t_i,m_h,c_k})}{\log(\underset{i,h,k}{\max}(U_{t_i,m_h,c_k}))} . \tag{5}$$

The reason for a logarithmic mapping is because in some cases the distribution of values for usages U is not uniform, and it may be that some very few high values for U_{t_i,m_h,c_k} might cause all the other usages to be invisible to the user because too much is concentrated in the lower part of the color scale model. Linear mapping and logarithmic mapping will be visually compared later.

2.4 Ordering of Subwindows

The next question to consider is the ordering of subwindows. Again, the main task of the application is not the analysis of correlation between usages in different courses. Rather we are interested in knowing the distribution of a module's usage in courses during a range of dates. We decided to order the subwindows following the natural order of the course IDs.

3 Graphical Representations

We implemented a prototype that shows the proposed approach. It consists of an application written in Java that runs in conjunction with the Moodle learning platform. The application remotely access the Moodle's database, and creates the graphical representations. Figure 2 illustrates the main GUI of this application.

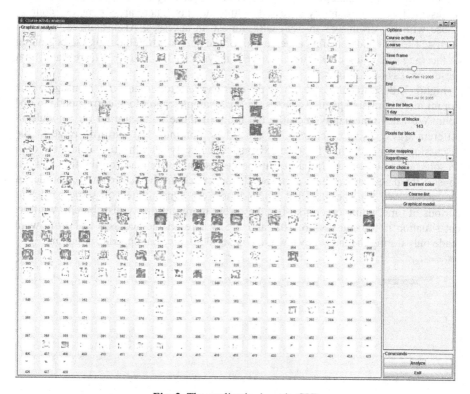

Fig. 2. The application's main GUI

As we can see from the picture, there are 2 main areas in the user interface. The panel on the right contains GUI components to receive the interactions from the user. In particular, the user can select begin and end of the time range, the granularity of the time discretization, the type of mapping (linear, logarithmic), and finally the color. He can also restrict the analysis to a subset of courses that can be selected using a selection window that appears when clicking on the button "Course list". The main panel on the left is dedicated to contain the matrix of subwindows. A label under the subwindow indicates the ID of the represented course. The image in Figure 2 represents 345 courses, which can be easily represented on a display with a resolution of 1280x1024 pixels. In this example, the user has selected a time range between Feb 13th, 2005 and Jul 6th, 2005 (which corresponds to the second semester of last year), daily granularity, and logarithmic mapping. The program automatically detected a viewable area of 143 pixels per subwindow, and a dimension of 9 (3^2) pixels per block. The result is a matrix of spirals that aim to help the administrator of the platform to make sense of the usage of a module in each course and compare the usage in different courses. The example in Figure 2 represents data for the module "course", which measures the attendance of the users to the course.

The picture allows managers, administrators, and instructional designers to gain useful insights. For example, it can be identified that only a small part of all courses represented were consistently accessed by the students. There are a bunch of courses (those with IDs from 343 to 372) with almost no accesses. Also, courses with IDs from 421 to 428 had very few accesses, all of them concentrated during the beginning of the range of dates considered. Further analysis on the course data revealed that these courses are still unused.

Figure 3 illustrates the case where the user has selected to analyze the usage of the Forum module in six specific courses. In this example each block corresponds to one hour. We can immediately see how the users of those courses used this module in different ways. Users of course 102 have done continue and uniform usage of the discussion forums. Users of course 284 have intensively used this tool either, but only on some specific periods. In particular, no discussions were made on the first and last days of the date range (which correspond respectively to the centre and the border of the spiral). We can see also that discussions were concentrated on some particular periods of time, while on course 102 they were more uniformly distributed. A different situation is for courses 66, 101, and 115, where all the accesses to Forum were made only during the latest days, while course 339 depicts an opposite situation.

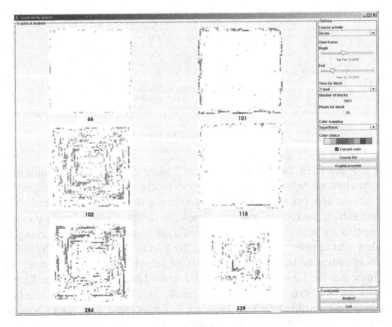

Fig. 3. Usage of discussion forums in six courses

Another example is represented in Figure 4. In order to have a more precise indication on the level of usage and to perform some comparisons, this example was set with a linear mapping. Each block is mapped with hourly time frames. We can see that courses 176 and 375 both had presumably two chat sessions on the whole period of the course. Course 202 instead had several chat sessions, but the blocks' color here is very light if compared to the chat sessions in courses 176 and 375. This could be explained by a low number of students that participated in chats in course 202, or a higher number of discussions that were made in courses 176 and 375.

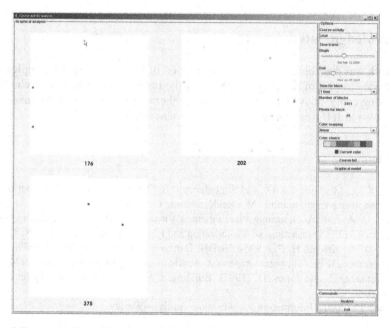

Fig. 4. Representation of the usage of chat with linear mapping for three specific courses

4 Conclusions and Future Works

We implemented a graphical tool that allows administrators and instructional designers monitoring the usage of modules in CMS. Thanks to an approach that uses the "pixelization paradigm", the tool allows to represent in a compact, single window mode, data on the usage of a module for several courses managed by a CMS. The tool adapts the visualization taking into consideration the display resolution, the number of courses to display, the discretization granularity selected by the user, and the date range. The result is a matrix of subwindows that map temporal data into blocks of pixels that follow a spiral placement model. The color (in particular, the saturation) of the blocks is mapped with the usage of the module made in the particular slice of time denoted by the block. The resulting visual representation may help the administrator of the platform and the organization's instructional designers to make sense of the usage of a module in each course and compare the usage in different courses.

We believe that these representations might also be useful to the instructors of the courses. For example, an instructor interested in knowing and comparing the usage of discussions in his course in different years might use this tool. In such case we should implement some security controls that doesn't allow an instructor to see courses owned by others. Moreover, the program can also be improved by allowing administrators, instructional designers, and (if needed) instructors to access the visualizations directly from the Moodle interfaces. A solution that uses a java applet wouldn't require the installation of another piece of software, while also allowing access to the visualization with a common web browser.

Acknowledgments

This work was possible thanks to the contribution of Mauro Nidola, who implemented the software as part of his Diploma degree at the Institute of Innovative Technologies of the University of Applied Sciences of Southern Switzerland in September 2005. Thanks also to Jeff Rose for proofreading the manuscript.

References

1. Card, K. S., Mackinlay, J. D., and Shneiderman, B. (1999). Readings in Information Visualization, using vision to think. Morgan Kaufmann, Cal. USA.
2. Keim, D. A. (2000). Designing Pixel-Oriented Visualization Techniques: Theory and Applications. IEEE Transactions of Visualization and Computer Graphics, Vol. 6, N. 1.
3. Keim D. A., Kriegel H. P. (1994). VisDB: Database Exploration Using Multidimensional Visualization. IEEE Computer Graphics & Applications, pp. 40-49. September 1994.
4. McCormack, C. and Jones, D. (1997). Building a Web-Based Education System. Wiley, New York.
5. Spence, R. (2001). Information Visualisation. Addison-Wesley.

Visu and Xtms: Point Process Visualisation and Analysis Tools

Jean-François Vibert[1,2], Fabián Alvarez[2], and José Pedro Segundo[3]

[1] Université Pierre et Marie Curie-Paris6, UMR S 707, Paris, F-75012 France ;
AP-HP, hôpital St Antoine, Service de Physiologie, Paris, F-75012 France
[2] INSERM, UMR S 707, ESIM, Paris, F-75012 France
jean-francois.vibert@upmc.fr, fabian.alvarez@u707.jussieu.fr
[3] University of California, Los Angeles, Department of Neurobiology,
Los-Angeles, California, 90095-1763 USA
segundo@ucla.edu

Abstract. This paper presents two tools developed initially to study neural networks information processing, but usable in any field that requires analyzing the behaviour of large data samples. One of the tools (*visu*) is devoted to the visualisation of simultaneously behaving events in several ways, either statically or dynamically along time, and allows examining each component individually. The other tool (*xtms*) is devoted to the analysis of temporal point process using a large number of different analysis, either involving single events or the interaction of two or three events, free running or cyclic. Examples in the domain of neural networks are given.

1 Introduction

Temporal point processes are time series in which the points represent the times when individual events occur. These events should contain the relevant information suitable for analysis, *e.g.* patients incoming or leaving a hospital, individuals infected during epidemics, occurrences of earthquakes, passage of cars at a toll gate, etc [1]. Temporal point processes are frequently encountered in biological data. One of the best known is certainly the spike train produced by neurons. The times of occurrence of the spikes compose the point process. It conveys the full information since spikes are judged events whose amplitude and shape are ignored while the interspike time interval, the spike frequency (or rate) and their modulations are the relevant parameters. The underlying continuous variation of the neuron's membrane potential and other parameters -which produce the spikes- are neglected in this point process analysis. Point processes are typically represented as dots arranged in several ways according to the problem to analyze. However, point processes are generated by complex mechanisms and the latter's analysis ultimately is crucial so as to provide the insights required to understand the point process itself. Neurobiologists use this kind of analysis since a long time [2][3].

We present two tools, which are originally part of the computational neurobiology workstation XNBC [4], developed to simulate biological neural networks. *visu* and *xtms* can be used as separate programs to visualize or analyze any point process,

P.P. Lévy et al. (Eds.): VIEW 2006, LNCS 4370, pp. 173–182, 2007.
© Springer-Verlag Berlin Heidelberg 2007

which !earlier was used, for instance, to examine epidemic simulations [5]. In this presentation, we will take examples of their use in the Neurosciences.

2 Visu: The Visualization Tool

Visu was initially developed to visualize simultaneously the membrane potential evolution of several hundred neurons. This is interesting since it allows to understand the way neurons communicate and act on their neighbours. While in neural networks, only spikes are information vectors, it is nevertheless interesting to be able to look at the membrane potential variations leading to the spike firing. Colour coding of these variations allows to easily understanding the way the network either synchronizes or desynchronizes. To work the program needs a text file with the membrane potential value of each neuron at each time step (usually 1ms). Other files describing the way neurons are connected can be provided and allow more complex (but more useful) representations.

Nevertheless, *visu* can be used to represent any type of data were several entities behave simultaneously along time and need a global visualisation tool. For example, while initially build to visualize neural network behaviour, it was successfully used to visualize the synchronization of flue incidence epidemics in the 22 french regions as observed by the French Sentinelles network.

2.1 Available Representations

Four representations are available. They are accessible either using the Option/Type of representation menu, or more directly by pressing one of the four pushbuttons in the menu bar. These representations are:

Linear Representation. This representation is like a dot display but shows as a colour graph the evolution of the membrane potential between spikes (Fig. 1). Each colour line represents the temporal evolution of the membrane potential of a neuron. Each column represents one simulation iteration. If several networks, a black lines delimit them. When a spike is fired, a red mark is drawn. With this representation, it is also possible to select one of the neurons with the mouse and by depressing the right mouse button to have a temporal representation of the intracelllular recording of the selected unit (Fig. 1). The time scale can be zoomed at will by selecting with the middle mouse button 2 limits. The right mouse button goes back to the original scale.

Intracellular Activity Representation. This representation displays simultaneously the temporal evolution of the membrane potential of selected units.

Global Activity Representation. This representation displays the temporal evolution of the number of units simultaneously spiking, and thus the global activity of the network.

Matrix Representation. This representation displays the network state at each iteration. The network is represented like a matrix. Each coloured square represents a neuron. Each black big square represents a cluster. Small lines represent connections from a neuron to another. When a neuron fires the spike can be seen travelling along its axon toward others neurons (Fig. 2). This representation gives information about the way spikes synchronize and how the conduction delay can influence the information processing. Matrix representation is only available if a network description is provided.

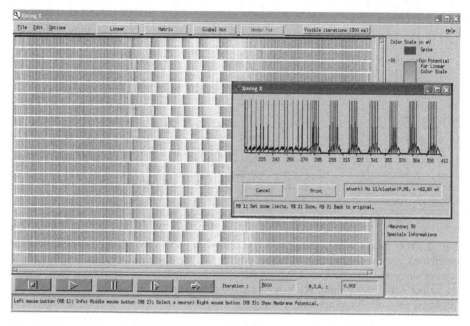

Fig. 1. The colour panel shows a screen of *visu* displaying the temporal evolution of the membrane potential of all the neurons of a simulated network. The colour coding scheme represents spikes (namely the point process) as red dots. The smaller panel inside shows the temporal evolution of one of the neurons selected using the mouse.

Colour Scale. The scale associates a colour to a value of the membrane potential. This scale has two parts: The first one (red rectangle) indicates the potential for a spike. The second one indicates a linear correspondence between colour and potential values. The top value can be modified (the bottom value is read in the *.def* file as the Negative Reversal Potential). By increasing this value, small variations of the potential can be filtered, and on the opposite by decreasing it, small variations of the potential around the resting potential can be magnified.

Time Step. The time between representations of each step of the simulation can be modified. This time acts on the visualization speed not on the simulation time step increment. The major purpose of this function is to slow down the visualization when the network is small or the computer too rapid.

Visualization buttons. Five buttons allow controlling the visualization like a video tape player. The button at left stops the visualization and returns to the first iteration.

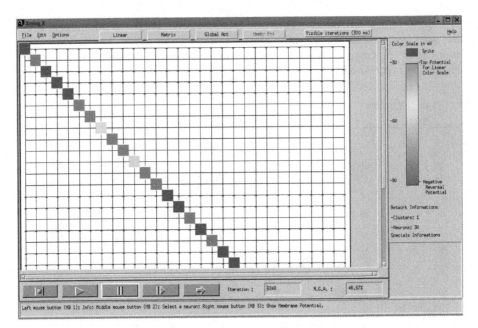

Fig. 2. The matrix representation of *visu* displaying the connections between neurons distributed into two interconnected networks. Neurons are represented by the small coloured squares along the diagonal. Colour is proportional to the membrane potential value according to the colour scale on the right. Black lines are the axons. Spikes are small red dots travelling clockwise on axons (same network as in Fig. 1).

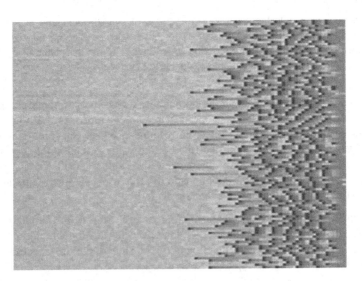

Fig. 3. The colour panel shows a *visu* display as in Fig.1 of the part of a network of 150 excitatory neurons sparsely connected and receiving noise. The colour scale is as in Fig. 1. Red dots are spikes. Output printed using *visu*.

The second starts the visualization. The third pauses the visualization at its current iteration. The fourth allows a step by step visualization. The fifth allows to go directly to the iteration number indicated the iteration text field at right.

2.2 Example of Visu Representation

Figure 3 display an example of representation of the activity of a randomly and sparsely connected excitatory network of neurons where noise induces by chance the discharge of a neuron, that in turn activates, thanks to sparse excitatory connections, some other neurons, that in turn fire others, etc. until the whole network almost synchronizes and becomes into a refractory state (in blue on Fig. 3), before coming back in a normal state, in which noise will be able to restart the phenomenon. This could be the way the respiratory rhythm is produced at the level of the brainstem.

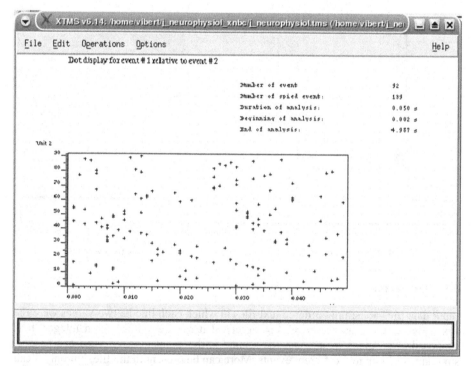

Fig. 4. The *xtms* window displaying the dot display of intervals for one studied event relative to a second test event. Each test event represents the zero and starts a new line on which the date of each studied event is plotted (during 500ms).

3 xtms: The Point Process Analysis Tool

Xtms is a tool devoted to the analysis of time series of point process. XTMS stands for Xwindow TiMe Series analysis. It provides principally dot display representations in order to put in evidence neurons synchronization, neuron response when a stimulus is given, etc. Point processes can be analyzed in more sophisticated ways, where pixel representation can help to reveal hidden phenomena.

Xtms allows making time series analysis from files storing events as point process. The default input file extension is .tms, the ouput is both on screen and as a PostScript file.

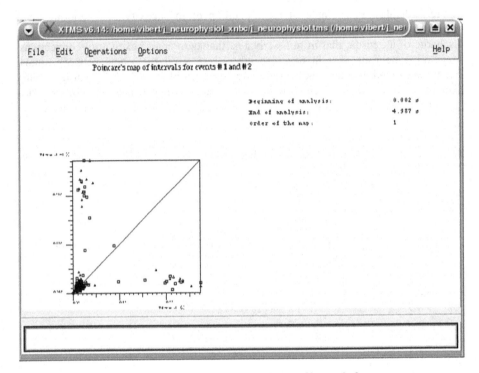

Fig. 5. *Xtms* window displaying the Poincaré map of intervals for two events

3.1 Data format

Input data files are simple ordered text files in which each line ha two values were the first value is the time (in increasing order), that can be a float or an integer. It is supposed to be given in ms, and the second is the event number. *xtms* can process simultaneously up to 5 different events. More can be present in the file, but only 5 can be chosen among them, which is largely enough, since only 1 to 3 are relevant for each processing. Only the data corresponding to the chosen events are stored. Data can be filtered; axis scales and several parameters can be chosen using specific menus. Up to 50,000 such pairs can be stored and processed.

3.2 Available Data Processing

Thirty-one different data analyses are available. In order to allow the processing of a series of graph menus of the main window are 'Tear-off' menus. This means that the user can always display all groups of functions menu on the screen.

The main menu propose some general actions and the following submenus : f(Time) drawings, f(Rank) drawings, Correlation analysis, Poincaré Maps, Inter and Peri Event Analysis, 3D drawings, Phase map analysis [dx/dt = f(x)], corresponding

Table 1. Analyses proposed by *xtms* with the number of events concerned and the output drawing type

Type of analysis	Event number	Output
F(Time) drawings		
Instantaneous rate = f(time)	1	line drawing
Instantaneous rate = f(time)	1	dots drawing
Instantaneous rate = f(time)	2	dots drawing
Rate as function of running time	1	histogram
Interval = f(time)	1	dots drawing
Interval = f(time)	2	dots drawing
Phase = f(time)	2	dots drawing
Phase = f(interval)	2	dots drawing
F(Rank) drawings		
Interval = f(rank)	1	dots drawing
Phase = f(rank)	2	dots drawing
Power spectrum of rank	1	histogram
Correlation analysis		
Autocorrelation	1	histogram
Crosscorrelation	2	histogram
Poincaré Maps		
Poincaré map of intervals	1	dot drawing
2 Poincaré maps of intervals	2	dot drawing
Poincaré map of phases	2	dot drawing
Poincaré map of cyclic data	2 or 3	dot drawing
Inter and Peri Event Analysis		
Inter event histogram	1	histogram
Dot display	2	dot drawing
Lissajous of cyclic data	2	dot drawing
Post event histogram	2 or 3	histogram
Post event interval pooled	2 or 3	dot drawing
Post event rate pooled	2 or 3	dot drawing
Post event phase pooled	3	dot drawing
Scatter diagram	3	dot drawing
3D drawings		
Interval I1 = f(I2,I3)	1	dot drawing
Phase P1 = f(P2,P3)	2	dot drawing
Phase map analysis [dx/dt = f(x)]		
D(interval)/dt = f(Interval)	1	dot drawing
exp(d(interval)/dt) = f(Interval)	1	dot drawing
D(phase)/dt = f(phase)	2	dot drawing

Post event histogram for Unit 2 and Unit 1 relative to Unit 4

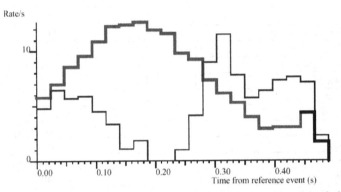

Pooled post event interval dot display for events # 2 and 1 relative to event # 4

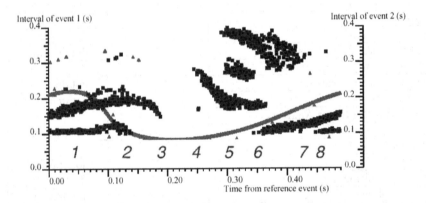

Fig. 6. Effect of high frequency steep modulation (sine: 0.490 s; 2.04 Hz; 34.694/s/s) given on an inhibitory fibre innervating a pacemaker neuron (crayfish stretch receptor). Top panel: cycle triggered histogram of presynaptic (red) and post-synaptic (blue) units. Bottom panel: Pooled cycle triggered dot display of interspike intervals of pre (red) and post (blue) unit's spikes. Portions with different forms are, for example, up to 1 (1:1 sliding alternation), 1–2 (windowing and intermittency), 2–3 (windowing and what appears to be 6:1 sliding alteration), 5–6 (intermittency), and 6–7 (messy erratic). Outputs printed using *xtms*.

to the 8 main analysis categories (Fig. 4 and 5). The detail of the proposed analysis is given in Table 1. For each analysis, specific questions are asked. All questions admit a default answer indicated in the text fields.

The computation functions need parameters to perform. Some parameters are common or generally used. These parameters can be set for global use, and act as default values already indicated in the panels presented to input analysis specific parameters. For each function the parameters that are not used are greyed and insensitive in the window. When started, and after a data file was read, *xtms* displays all the present events in order to select those to analyze. Then all fields that can be deducted from the file are filled, and can be modified (for example begin and end of the processing). Events to select are the analyzed event or a stimulus, a clock or another event. Output is always a graph, a histogram, a plot or a dot display. Individual values of histograms can be saved in files for other processing. Some descriptive statistics can be produced, and in this case, are drawn on the graph.

3.3 Interest in Revealing Hidden Behaviours

Dot displays are common use in neurophysiologic data analysis since point process time series are common in this field. Biological rhythms are also common, but not always easy to analyze, due to cycle variations, leading to use averaging or statistical analysis. Nevertheless this kind of analysis where data are individually analysed along with others can reveal behaviours hidden by averaging or global visualization. An example is shown on Fig. 6, were the effect of high frequency steep modulation is sent to an inhibitory fibre with a synapse on a pacemaker neuron (a crayfish stretch receptor). As expected, the cycle triggered histogram of the discharge frequency of both pre and post synaptic fibres are opposite in phase, but with irregularities. Using pooled cycle triggered dot display of interspike intervals of pre and post unit's spikes, in opposite, shows that strange attractors exist that can explain these irregularities. Dot display representation reveals nested information about the coding and transfer function of the synapse, and shapes of the pixel representation differ in irregularity and predictability leading to describe different types of neural coding of the synapse, according to the driving frequency, amplitude, slope, etc. (for a full description of these phenomenon see [6]).

4 Tools Flexibility

These tools were developed as a part of a large project in computational neuroscience, but they can be used for other purposes if the correctly formatted input data files are provided. These files are ascii files easy to produce with other programs.

For *visu* the necessary file is the file containing the unit numbers and the membrane potential values at each iteration. Other files are loaded (network organisation and links between neurons), but they are not mandatory (only some functionalities will be nor available such as the representation of Fig. 2). The number of neurons that can be visualized with reasonable comfort is depending on the computer memory and speed, since the necessary memory is dynamically allocated as needed. Adaptation to other type of data, such data with parallel evolution but not necessarily the same interval between samplings, they can be represented by building the data file by filling the unavailable data with the same value for each time step, using the smallest time step of the time series.

For *xtms* only one file is necessary, also an ascii file easy to produce. The number of data that can be analyzed is here again depending the computer configuration. It exist other tools in the open source domain. One of these is *RasterViewer* [7]. It was developed for the computational neuroscience domain, as *xtms*. *RasterViewer* is complementary to *xtms*, and was included in the recent versions of XNBC [8].

5 Conclusion

We describe here programs allowing the representation and analysis of point process time series, stored as simple time ordered ASCII files, with several events embedded, some of them being possible cycle markers, stimuli, etc. The programs allow selecting a time range and data filtering, and produce several kinds of visualisation and outputs, all involving physiologically relevant facets. These programs are written in C, are GPL open source and run under Linux or Windows. They are available in binary for Linux and Windows and in source code. These tools need the presence of an X11 server, always found on Linux, not on Windows. For this reason, Xming [8] is distributed simultaneously to the Windows version of XNBC [9] were these tools can be found.

References

1. Brillinger D. R. Time Series: Data Analysis and Theory. Holt, Rinehart and Winston, Inc. Pub, New-York. (1975) 500pp
2. Moore GP, Perkel DH, Segundo JP. Statistical analysis and functional interpretation of neuronal spike data. Annu Rev Physiol. 28 (1966) 493-522.
3. Villa, A. E., Tetko, I. V., Iglesias, J.,. Computer assisted neurophysiological analysis of cell assemblies activity. Neurocomputing 38-40 (2001)1025–1030.
4. Vibert J-F, Alvarez F., Kosmidis EK. XNBC V9: A user friendly simulation and analysis tool for neurobiologists. Neurocomputing 38-40 (2001) 1715-1723
5. Toubiana L., Vibert J-F. A neural network model for the spread of communicable diseases. In: Gierl L., Cliff A.D., Valleron A.-J., Farrington P., Bull M. (Ed.) Proceedings of GEOMED'97. Tuber-Verlag Ed, Berlin, (1998) 249-259
6. Segundo JP, Vibert JF, Stiber M. Periodically-modulated inhibition of living pacemaker neurons--III. The heterogeneity of the postsynaptic spike trains, and how control parameters affect it. Neuroscience. 87 (1998) 15-47.
7. Iglesias J. Emergence of Oriented Circuits driven by Synaptic Pruning associated with Spike-Timing-Dependent Plasticity (STDP). Thèse de doctorat, Universités de Lausanne (Suisse) et Grenoble (France). (2005) 162pp
8. XNBC: http://sourceforge.net/projects/xnbc/ and http://www.b3e.jussieu.fr/xnbc
9. Xming: http://wiki.freedesktop.org/wiki/Xming

Visualizing Time-Course and Efficacy of In-Vivo Measurements of Uterine EMG Signals in Sheep

Gaj Vidmar, Branimir L. Leskošek, and Drago Rudel

University of Ljubljana, Faculty of Medicine, Institute of Biomedical Informatics
Vrazov trg 2, SI-1000 Ljubljana, Slovenia
{gaj.vidmar,brane.leskosek,drago.rudel}@mf.uni-lj.si
http://www.mf.uni-lj.si/ibmi-english

Abstract. A method for constructing condensed yet concise visualization of a long-term bioinformatics research project is presented. The chart combines information on data quality and chronology of research activities. It is implemented as Microsoft® Excel spreadsheet with VBA automation. The project comprised 3100 hours of EMG measurements in 23 pregnant sheep, each monitored for up to 5 months during pregnancy and immediately after labor. EMG signals were recorded from the surface of the uterine wall by electrodes implanted at the horn and cervix. The signals were assessed for quality, stored in electronic EMG archive and queried from there for visualization. The spreadsheet shows subject data as time-sorted column headings and pregnancy time along rows, with cell color depicting measurement efficacy rating (signal unusable, partly usable, OK), and a symbol marking labor date. The chart is best read if rotated into landscape format. Project group members confirmed usefulness of the chart in assessing research progress and spotting measurement problems.

1 Introduction

In the information society, the need for highly condensed yet accurate presentation of large amounts of categorized information for decision making purposes is ever increasing. The most inventive and successful methods and tools are both grounded in computer technology progress and inspired by the challenges it brings, like Treemap [1, 2], Holographic Research Strategy [3], or sparklines [4]. Such visualization often incorporates the time dimension, which is crucial in project management [5] or in the financial markets (www.panopticon.com). It is also important in other areas where a clear picture of complex system dynamics can improve the course of action, e.g., when monitoring hohspital activity via case mix [6]. Scientific projects in the field of biomedical informatics that involve years of high-tech, financially demanding and time-consuming measurements are another prime example. Hence, we aimed at constructing such a display as a tool for helping a long-term research project on electromyographic (EMG) activity of the uterus in pregnant sheep follow the success of the measurements.

P.P. Lévy et al. (Eds.): VIEW 2006, LNCS 4370, pp. 183–188, 2007.

The whole project involved 35 sheep; the measurements addressed in the paper involved 28 pregnancies in 23 normally gravid sheep. All were multiparas with one or two fetuses. At low gestation (usually 30-60 days), each sheep was instrumented with EMG measuring equipment [7]. A pair of platinum electrodes was surgically attached to the uterine wall surface at the pregnant uterine horn, and the other pair at the cervix. The instrumented sheep were located in a stall with normal living conditions, where they moved around freely, even during measurements. Namely, two weeks after electrode implantation and veterinary care, the sheep were equipped with external EMG pre-amplifier worn in a belt around the body. During measurement, the pre-amplifier was wired to portable measuring equipment outside the stall.

As a rule, EMG measurements were performed at least twice a day, in the morning and in the afternoon, before labor, during labor and one week postpartum. In total, over 3100 hours of EMG recordings were made. A typical EMG recording session lasted for one hour during day, and for ten hours over night. The number of measurements per day in a sheep ranged from one to eleven, with few sheep subject to five or more sessions in a single day. The period for which a sheep was included in the research ranged from 3 to 134 days (median 74 days).

Two channels of EMG activity were recorded, sampled at 20 Hz frequency. The digitalized EMG signals were stored in the electronic archive database. To study various research hypotheses, the EMG data were mathematically processed in time, frequency and time/frequency domain. Subsequently, aggregated signal parameters were statistically analyzed.

Based on signal processing over one-minute intervals, each measurement session was visually assessed and classified into one of three categories:

- 0 – Whole record had to be removed from further analysis;
- 1 – Parts of the record had to be rejected;
- 2 – The whole record is OK.

In total, 3488 sessions were considered. For the horn, 66.9% of the sessions were rated as OK, 3.6% as partly usable, and 29.4% as unusable. For the cervix, the success rate was better: 75.6% of ratings were 2, 2.4% were 1, and 21.9% were 0. The main reason for an EMG measurement session to fail was damage made to the electrode wires externally by the sheep.

2 Method

The chart is produced as Microsoft® Excel spreadsheet cell-chart with VBA automation. It requires the data to be placed on one sheet (in adjacent columns: subject name, measurement number, date, horn measurement efficacy rating, and cervix measurement efficacy rating; in the final column, the first row of data for each subject contains labor date). The chart is produced as a separate worksheet with gridlines hidden. The required data can be directly queried from our electronic EMG archive. The macro requests the user to input one parameter, the number of cells that represent one day (from 2 to 168, the default is 12). For our data, the default value of 12 was also the most suitable value in terms of trade-off between totally undistorted depiction (which would be ensured by divisibleness of the number by each encountered number of measurement sessions per day) and higher data-ink ratio [4] of the chart (which decreases with the number of cells per day).

The first two rows of the resulting worksheet are column headings with subject data (sheep name, date of labor, channel locations); the number of columns for each sheep corresponds to the number of channels used, and a narrow blank column separates the sheep. Within each channel column, each day (from the first to the last day of sheep's measurement period) is represented by a box: if there was no measurement on a given day, the box has neither border nor fill, so it is not drawn on the final chart; otherwise, the bordered box is vertically divided by the number of measurements per day (which is easily achieved by applying borders to worksheet cells given a sensibly chosen number of cells representing one day). Each box, i.e., pixel in terms of the pixelization paradigm [6, 8], is colored according to the measurement efficacy rating (0 – grey, 1 – yellow, 2 – white). A symbol in the empty column marks sheep's labor, so that the vertical time scale in days is anchored.

The final form of the chart is produced with the Microsoft® Excel's page setup feature by specifying that the document printout should be 1 page long and 1 page wide. For our dataset (with 12 cells per day), this just worked, but for bigger datasets, two-page long (or wide, or larger) charts should be produced (our VBA macro is limited just by maximum worksheet size). The chart can be conveniently exported for publishing or further processing by printing it to a PostScript printer assigned to a file (or using one of the many similar solutions). Readability is better if the chart is viewed in landscape format, so that subjects are represented by vertically stacked rows with time running in the horizontal direction, and in further discussion of the chart we refer to that orientation.

3 Results

The chart is presented in Fig. 1. From the header, we can read that the earliest measurement took place on 30 June, 1995, with sheep B, and that the last sheep included in the study was Ajda (her first measurement took place on 6 January, 2005). Next to the header, the letters H and C are put for the first and the last sheep, denoting horn and cervix measurements, respectively. Since the same pair of row headers applies to each sheep, the letters are omitted in-between for clarity.

The labor-denoting red arrow-like symbols placed at the end of the green vertical lines show that in three sheep, no labor took place. For F, Kaja (first pregnancy) and Mimi (first pregnancy), pregnancy must have terminated at some point in spontaneous abortion without the research staff or the farmer keeping the stall noticing, hence the cells with starting date of measurement for these sheep are colored grey rather than green. In the majority of the other sheep, the increased frequency and shorter duration of measurement sessions around labor, which was part of the study design, is clearly visible. Another observation quickly offered by the chart is that no successful measurement was obtained in only three pregnancies. A general message conveyed by the chart is that during the first half of the project, the follow-up periods were increasing, while the most important overall observation is that the share of completely successful measurement sessions increased over the course of the project.

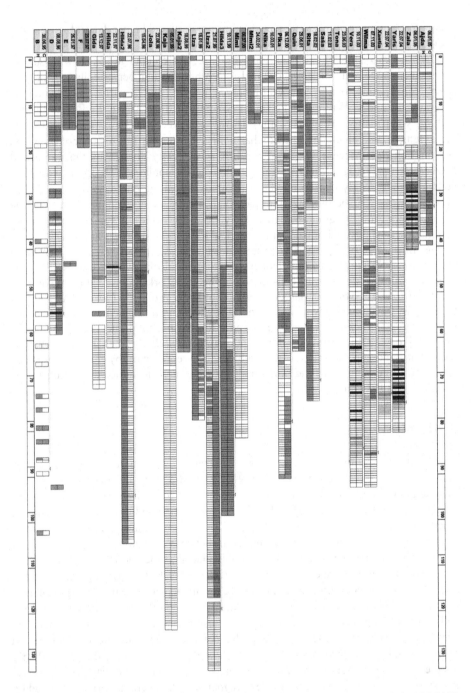

Fig. 1. Uterine EMG measurements in pregnant sheep over research project period 1995-2005. Sheep names run from B (first measured on 30 June, 1995) to Ajda (6 January, 2005); H= horn, C=cervix; red arrow shows date and time of labor (see Sect. 2 for details).

4 Discussion

Our charting procedure is based on the Microsoft® Office suite on the Windows platform. As an alternative, a dedicated application for producing the chart could be developed, or it could be implemented in a cross-platform environment, e.g., as a Java™ applet, or in R [9]. Our spreadsheet-based implementation has natural limitations arising mainly from the issues of printer page size and resolution, but despite its relative simplicity it produces complex charts conveying large amount of information with minimum manual intervention from the user in just a few seconds on a standard personal computer.

The research-project group members confirmed the usefulness of the chart for displaying the quality of EMG recordings and thus presenting the success of their measurement efforts in general. The chart enabled them to spot in which sheep and in what time periods measurement problems occurred. Of course, such benefits cannot be quantified, but the problem spotting markedly aided in planning further sheep instrumentation and improving measurement set-up, and it is evident that the percentage of successful EMG measurement sessions subsequently increased. While the proposed visualization is primarily a tool for aiding research rather than a result of research, the chart for the present state of the whole EMG archive provides a summary of overall project progress that is useful and interesting in its own right.

In future research of uterine EMG, which is aimed at benefiting pregnant women in the long run, the chart's design principle could be incorporated into a graphic account of status of a single patient [10] or a group of patients. The presented visualization method can be applied to many fields of medical informatics, and in that sense, as well as in its design principles, it is similar to the generalized Caseview [8]. Namely, like the Caseview approach, it defines a reference frame based on three criteria: one nominal variable that can be meaningfully ordered (subject, sorted by date of entry into the study), one discrete or discretized dimension (time, which is treated as a ratio scale rather than just an ordinal one that is required by Caseview), and one binary attribute (used for splitting the frame in Caseview, but in our case it is measurement location and it defines row pairs). A further similarity with Caseview is ordinal color scale applied to the cells, while instead of numbers in cells used in Caseview, additional information is depicted with labor markers.

In conclusion, we would like to stress that the presented method should be used together with other visualization methods and amended whenever needed. The nature of contemporary data visualization is dynamic – in the words of one of its pioneers, Jacques Bertin [11]: "*A graphic is no longer 'drawn' once and for all: it is constructed and reconstructed (manipulated) ... A graphic is never an end in itself: it is a moment in the process of decision making*".

References

1. Shneiderman, B.: Tree visualization with tree-maps: 2-d space-filling approach. *ACM Transactions on Graphics (TOG)*, 11(1) 92–99, 1992.
2. Bederson, B.B., Shneiderman, B., Wattenberg, M.: Ordered and quantum Treemaps: making effective use of 2D space to display hierarchies. *ACM Transactions on Graphics (TOG)*, 21(4) 833–854, 2002.

3. Végvári, A., Tompos, A., Göbölös, S., Margitfalvi, J.: Holographic research strategy for catalyst library design. Description of a new powerful optimisation method. *Catalysis Today*, 81(3) 517–527, 2003.
4. Tufte, E.: *Beautiful evidence*. Graphics Press, Chesire, CT (2005, in press).
5. Plaisant, C., Chintalapani, G., Lukehart, C., Schiro, D., Ryan, J.: Using visualization tools to gain insight into your data. In: *SPE Annual Technical Conference and Exhibition, 5-8 October, Denver, Colorado*. Richardson, TX, SPE (2003) 1–9.
6. Lévy, P.P.: The case view, a generic method of visualization of the case mix. *International Journal of Medical Informatics*, 73(9-10) 713–718, 2004.
7. Leskošek, B., Pajntar, M., Rudel, D.: Time/frequency analysis of the uterine EMG in pregnancy and parturition in sheep. In: Magjarevic, R. (ed.): *Biomedical measurement and instrumentation - BMI'98. Proceedings of the 8th international IMEKO TC-13 conference on measurement in clinical medicine and 12th international symposium on biomedical engineering, Vol. 3*. KoREMA, Zagreb (1998) 106–109.
8. Lévy, P.P., Duché, L., Darago, L., Dorléans, Y., Toubiana, L., Vibert, J.-F., Flahault, A.: ICPCview: visualizing the International Classification of Primary Care. In: Engelbrecht, R., et al. (eds.): *Connecting Medical Informatics and Bio-Informatics. Proceedings of MIE2005*. IOS Press, Amsterdam (2005) 623–628.
9. R Development Core Team: *R: A language and environment for statistical computing*. R Foundation for Statistical Computing, Vienna, Austria (2005). ISBN 3-900051-07-0, URL http://www.R-project.org
10. Powsner, S.M., Tufte, E.R.:Graphical summary of patient status. *The Lancet*, 344(8919) 386–389, 1994.
11. Bertin, J.: *Graphics and graphic information-processing*. de Gruyter, New York (1981).

From Endoscopic Imaging and Knowledge
to Semantic Formal Images

C. Le Guillou[1,5], J.-M. Cauvin[2,5], B. Solaiman[4,5], M. Robaszkiewicz[3,5], and C. Roux[4,5]

[1] UFR Médecine, Université de Bretagne Occidentale, CHU Morvan, 5 Avenue Foch,
29609 Brest Cedex, France
Clara.LeGuillou@enst-bretagne.fr, Clara.LeGuillou@univ-brest.fr
[2] Département d'Information Médicale - [3] Service d'Hépato-gastro-entérologie,
CHU Cavale Blanche, Boulevard Tanguy Prigent , 29200 Brest, France
{Jean-Michel.Cauvin, Michel.Robaszkiewicz}@chu-brest.fr
[4] Département Image et Traitement de l'Information, École Nationale Supérieure des
Télécommunications de Bretagne, Technopôle Brest-Iroise, 29238 Brest Cedex 3, France
{Basel.Solaiman, Christian.Roux}@enst-bretagne.fr
[5] Laboratoire de Traitement de l'Information Médicale, INSERM U650, 5 Avenue Foch,
29609 Brest Cedex, France

Abstract. Provided with evolved functionalities, a digestive endoscopy atlas can be used as a tool of training and even of diagnosis aid. The architecture of such a system is leaning on two bases, one of endoscopic knowledge another of case iconography. Being inspired by medical practice, a bi-leveled – disease knowledge base allows a classification of possible diagnoses and a case selection of the endoscopic case base, enabling the similarity step to complete the retrieval. This project benefits at many levels from the "pixelization paradigm". Indeed, to visualize the Knowledge and Case bases is of great interest, but it's more exciting to visualize the steps of the classification and of the similar case retrieval by the generation of images confronting the knowledge base and the case base to the new case.

1 Introduction

Since its widespread use in the late 1960's, endoscopic examination of the upper gastrointestinal tract (esophagus, stomach and duodenum) has expanded the understanding of numerous gastrointestinal diseases thanks to a careful inspection of the mucosal surface and has thus greatly improved the ability to care affected patients (see Figure 1). When the endoscopist is exploring the digestive cavity, the action of focusing on a particular organ area and of evoking diagnostic hypotheses underlies a complex march of the thought. Indeed, Physician's reasoning [1] emphasizes, in the diagnostic process, two decision levels which refer to two information spaces: the endoscopical findings, i.e. the lesions, and the diseases. An interaction connects these two decision levels because the diagnosis of endoscopical findings meddles with the disease diagnostic decision. Inconsistencies in the final decision according to other information (medical context, other endoscopical findings) must lead to doubt about the validity of endoscopical finding diagnoses. At the disease level, the decision of assigning a

P.P. Lévy et al. (Eds.): VIEW 2006, LNCS 4370, pp. 189–201, 2007.
© Springer-Verlag Berlin Heidelberg 2007

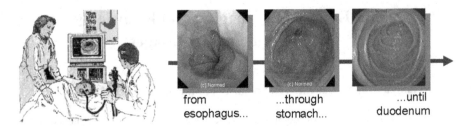

from
esophagus...

...through
stomach...

...until
duodenum

Fig. 1. The upper digestive video-endoscopic examination

diagnostic class also depends on the prevalence of this diagnosis in the current practice and on the endoscopist experience for rare cases.

Almost no research bends over this medical domain in order to promote the conception of decision aid tools. This challenge is meanwhile ours with a computerized advanced adaptation of medical atlas-books. Next to a classic illustration of diagnoses by means of noteworthy iconography, such an atlas must, in addition, integrate the similar case retrieval by the diagnostic hypothesis evaluation.

In general approach, for the present, there is no question of exploiting the numerical content of endoscopic images or sequences because such content would not be sufficient to "translate" the medical meaning of the situation. As the section 2 shows it, the followed approach consists in elaborating a model of the situations encountered with the imagery, that is to say a knowledge base of endoscopic pathologies exemplified by a case base, i.e. images and/or sequences acquired from endoscopic examinations, illustrated by their diagnostic reports, on one hand, and a set of independent images, on the other hand. The architecture of such a system is in keeping with the Pattern Recognition and Case Base Reasoning [3] paradigms, leaning on the two bases. The retrieval of similar cases can be assimilated to process with two steps, classification (global method) owing to the knowledge base and, afterwards, measurement (local method) by means of selected cases of the endoscopic case base. Lesion identification and Disease recognition – that is, the Scene analyze [2] - are the driving forces not only in the Classification stage, but also in the Similarity measure stage. In front of huge amount of data and of processing, the section 3 presents the methods used to facilitate their exploration and understanding, that is to say, their transformation in images. Thanks to the Pixelization Paradigm, scenes and objects images and even a knowledge base image can be produced as well as case images and globally an image of the case base. At the level of processing, the visualization of the steps is very useful in order to verify the classification and similarity measure algorithms of the similar case retrieval. The section 4 will conclude on the interests and limits of such system, on the pixelization contribution to ameliorate it and on the various ongoing evolutions.

2   Knowledge Representation and Reasoning in Endoscopy

Drawn from the Minimal Standard Terminology of the European Society of Gastro-Enterology (ESGE) [4], a two-leveled description mode of the endoscopic imaging and of the pathologies is illustrated by the concept of Scenes with Objects.

The relevant case retrieval is based on two approaches: a first step classifies owing to the knowledge base and, after, a similarity measure will complete the retrieval.

2.1 A Scene/Object Modeling

Owing to the federative concept of Scenes with Objects, the description of endoscopic information, as shown Figure 2, allows to distinguish three types of Scenes:

- **Physical Scenes**
  The file of an image or of an image sequence is considered to be a Physical Scene. It visualizes an interesting part of the endoscopic exam, showing anomalies, that is the Objects.

- **Logical Scenes**
  A Logical Scene represents a medical interpretation of endoscopic imagery, i.e. an endoscopic disease diagnosis, which associates a peculiar patient context, one or several endoscopic Finding(s) or Object(s) and their eventual spatial relations.

- **Conceptual Scenes**
  As abstractions of Logical Scenes, Conceptual Scenes are the extended definitions of the upper digestive tract pathologies. Patient context, reasons for the endoscopy, one or several Conceptual Objects with their eventual spatial relations, and the complementary procedures to be advised, constitute the medical knowledge of these Scenes.

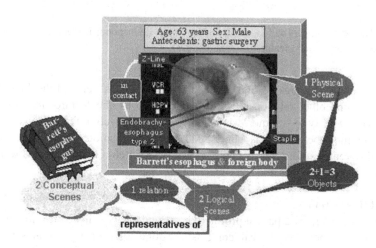

Fig. 2. Example of Scene Object modeling: the image (Physical Scene) of a patient with a peculiar clinical context, which shows three findings (Logical Objects), that represent two diseases (Logical Scenes); Conceptual Scenes denotes the knowledge associated to diseases

2.2 The Scene/Object Information

Object Information

Lesions or any element of interest, i.e. the "endoscopic findings", constitute the objects to be depicted thanks to an exhaustive description mode (see Figure 3). So, each object is described with 24 features (even 33 if a sub-object exists). To each feature is associated a set of choices, representative of all possibilities and judiciously defined by the expert.

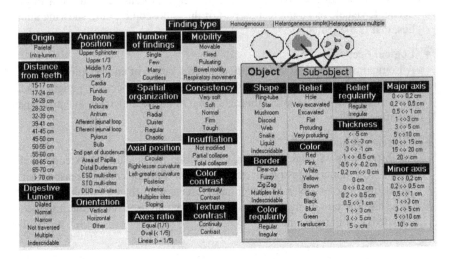

Fig. 3. The 24 features (145 modalities) of an Object description; if there is a sub-object, the description is made on 33 features (206 modalities)

As sub-object features depend of the "non-homogenous state" of the Type feature, there are some other relationships between modalities and feature (for example an object with Density to "unique" has not a Spatial Organization feature, an object with Shape to "ring-tube" has not a Minor Axis feature, and so on...) or between modalities of different features (for example, modalities of Relief and Thickness features or modalities of object sizes and sub-object sizes,...).

Scene Information

A scene is depicted by a patient profile (the sex and age prevalence features as well as a predefined whole of clinical contexts denoting antecedents, circumstances and symptoms), by the objects (at least one), by eventual spatial relations between objects and the complementary procedures to be envisaged to confirm the disease diagnosis.

Object. As shown on Figure 4, information on disease diagnosis well encompasses the four relevant items.

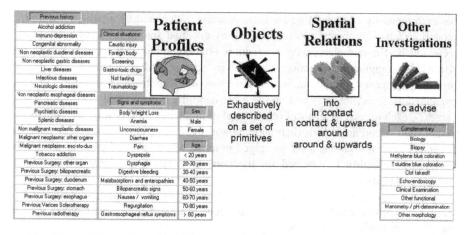

Fig. 4. Scene Information with the patient profile features and the complementary feature

2.3   The Knowledge and Case Bases

Endoscopic Knowledge Base

As an expert squeezes out his knowledge, using linguistic valuations and instilling even doubt, the chosen way (see Figure 5) is supposed to do the same, manipulating linguistic truth degrees as well as uncertainty or vagueness [5]. Each feature can be judged as without interest, impossible or of interest; in this last case, each feature item must be evaluated as *never, exceptional, rare, frequent* and *always*. A sixth level is also defined (*doubtful*) when the choice between never and exceptional cannot be stated: this level expresses the knowledge limits of the expert.

An object is systematically illustrated upon 206 modalities corresponding to the 33 features with the Knowledge Linguistic Valuation and is marked by ESGE code and libel. The expert gives each feature an Interest valuation and, valuations of Incidence

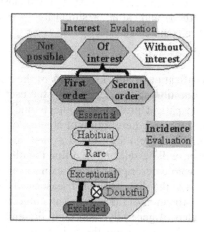

Fig. 5. Knowledge Linguistic Valuation (Incidence degrees and Interest degrees)

for each modality of interesting features. A scene is depicted by a patient profile (51 modalities), by the objects (86 modalities), by eventual spatial relations between objects and by the complementary procedures (10 modalities), all these still with the Knowledge Linguistic Valuation, is called by an ESGE denomination. The patient profile is composed, on the one hand, by the Sex and Age Prevalence features that are evaluated as object features and on the other hand, by a predefined whole of clinical contexts, each judged in terms of interest and of incidence. In this way, the objects are assessed in accordance with their interest and incidence, but one of them must be Essential, Spatial relations are also judged in terms of interest and of incidence. Only an interest valuation is required for the complementary procedures. Finally, each disease is classified as main or not.

To sum up, object and scene descriptions, form a two-leveled knowledge base constituted by approximately 150 Conceptual Scenes and 86 Conceptual Objects (that is the 86 object modalities of the Scenes).

Endoscopic Case Base
For the imagery, it is a whole of endoscopic examinations, which must be indexed. Constituted of images or sequences and of a diagnostic report, each examination represents a set of *Physical Scenes* as well as a set of *Logical Scenes*. While a *Physical Scene* is a file where are visualized the objects, a *Logical Scene* represents an endoscopic diagnosis which consists of information concerning the patient (common to all the *Logical Scenes* of a same examination) and of all the objects attached to the diagnosis. Moreover, the indexing must hold into account that *Logical* and *Physical Scenes* do not tally. The case base is constituted roughly by 70 examinations (constituted of images or sequences and of a diagnostic report) and 150 images.

2.4   Towards Solving New Cases

Usually, the similar case retrieval is based either on a Similarity measure (computational approach) or on an Indexing structure (representational approach). But here, these two approaches are to be combined: a first step will classify owing to the Indexing structure (global method) and, after, a similarity measure (local method) will complete the retrieval of relevant cases.

Classification step
Lesion and Disease Classification are the key points of the Scene analyze (see Figure 6). With this intent, the questioning interface allows the user to depict an endoscopic exam, in other words the patient profile, the objects and possible spatial relations between objects. To avoid a boring description of objects, only 5 features (anatomical position, form, color, relief and type) are wanted at first; the other ones, according to their discriminating power, will be selected to refine the object recognition. After the object classification, the whole description of the exam is analyzed to identify one or several Logical Scenes - i.e. the diagnoses of diseases -. This analysis of scene still allows perfecting the classification of the lesions - i.e. objects-. Indeed, as the level of the image (or of the sequence) generally represents the endoscopic lesion level, the classification should especially insist on the objects in order to select a subset of

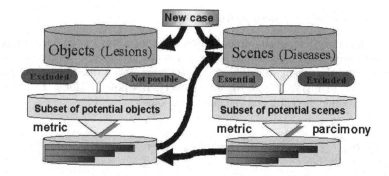

Fig. 6. Scene analyze: object classification next scene classification and then object sorting reassessment

candidate cases (the similarity playing upon those), while, from a medical point of view, the level of interest should surely envisage the disease diagnosis.

Similarity step
The classification step having unveiled the objects and scenes potentially similar, it is a matter of illustrate them by pertinent images. For this purpose, the new case is compared with selected cases according to three levels, modalities, objects, and scenes. The Modality Similarity uses predefined cross tables for each feature, where the comparison between two modalities is judged with 5 values (incompatible - no similar - slightly similar - fairly similar - identical). Fusion of Object Similarities, of patient profile similarities and relation similarities complete the retrieval.

3 Pixelization in Endoscopy

From 1987, a report [6] recommended that the National Science Foundation subsidize a longer-term research in what is now known as the field of scientific visualization where calculations require visualization to present their output in an understandable form. In contrast, Information Visualization aims at an interactive visualization of abstract non-spatial phenomena such as bibliographic data sets [7], web access patterns, etc [8]. Indeed, the intention of information visualization is to use our perceptual and visual-thinking ability in dealing with phenomena that might not readily lend themselves to visual-spatial representations [9]. Advances of information visualization were significantly driven by information retrieval research [10]. With the Geographic Information Systems (GIS), information visualization employs 2D methods. Geographic coordinates provide a most convenient and natural organizing framework to accommodate a wide variety of information [11]. This 2D approach refers to the "Pixelization paradigm" where a pixel of an image denotes a data.

All the descriptions of endoscopic information and of processing, previously presented, deserve a pixelization that illustrates the "Transformation of the symbolic into the geometric" [6].

3.1 Endoscopic Information Pixelization

With the objects systematically described on 206 modalities by the Knowledge Linguistic Valuation, it is evident that a pixel denotes an object characteristic and its color represents the expert valuation. An object represents a line of pixels (figures 7) and altogether an image (figure 8). Interactively, it's possible for a given pixel to know what finding, modality and valuation it denotes.

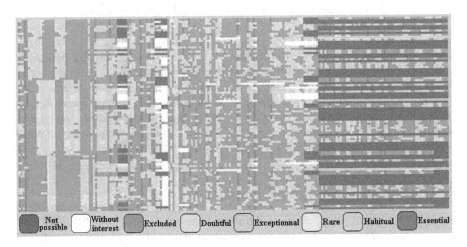

Fig. 7. Visualization of Tumor descriptions (esophagus, stomach, duodenum)

Fig. 8. Knowledge Base – Object level – (86 objects x 206 modalities)

Likewise, the 148 Scene descriptions of the Knowledge Base can be visualized as shown figure 9. Each row represents a Scene with on one hand, 62 columns denoting the modalities regarding the clinical contexts (age, sex, antecedents, circumstances and symptoms) and complementary investigations, and, on the other hand, 86 columns denoting the valuations the objects (which are the rows at the object level).

Regarding the cases, the pixelization approach is a bit different because based on the three layers of an RGB image. A case object is represented by three vectors of 206 components. At each level corresponds an information type as seen figure 10. The Object description vector represents the Logical Object. Each of its components corresponds to one modality. Items which describe the object are set to 1 and the other ones to 0. The Object knowledge vector traduces Conceptual Object. Each of its components is associated to one item, and takes its value between 0 and 1 representing the Linguistic Valuation from *"excluded"* to *"essential"*. The third vector

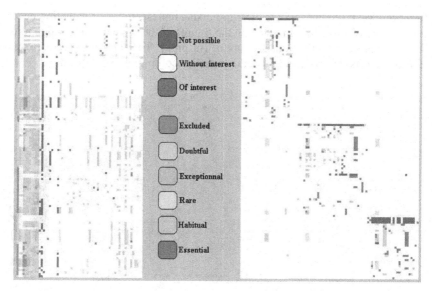

Fig. 9. Knowledge Base - Scene level – (148 scenes x 62 modalities) and (148 scenes x 86 objects)

expresses the similarity valuations centered on case object description. Each value between 0 and 1, from *"incompatible"* to *"identical"*, traduces the comparison between a modality and the described modality.

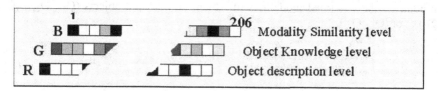

Fig. 10. Case pixelization - Object level –

This case representation allows introducing the information useful in order to solve other cases. Indeed, with the knowledge level, the cases of same diagnosis can be retrieved and with the similarity level, the visual aspect is explored, suggesting cases with similar or differential diagnoses. At the level scene of the cases, the same approach fit the bill.

3.2 Classification Pixelization

As shown figure 11, the first step of the object classification consists in adapting the object knowledge base. For example, if the new described case is with the feature type as homogenous, all sub-object features are impossible and with the feature density as single, organization feature is also impossible, etc… The following step selects a sub-image of the object knowledge base with only, in column, the modalities of the description. Next, the lines where there are pixels representative of valuation never or impossible are disqualified (in black on the figure). If there are doubtful valuations on a line, the object classification is doubtful (in pink). The lines (in green) represent potentials objects retrieved and there are ranked according to their typicality.

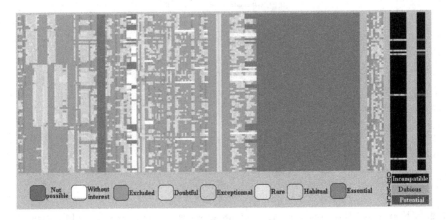

Fig. 11. Object classification visualization

The scene classification is more complicated as the scene described is complex. For example, a scene with two objects, one with two potential diagnoses O_{11} and O_{12}, the other with four $O_{21},.. ,O_{24}$, implicates a combinational process of eight scene classifications. The figure 12 shows the selected sub-images of the knowledge base with in columns the described modalities of which the potential objects (O_{11}, O_{21}), (O_{11}, O_{22}),...or (O_{12}, O_{24}).

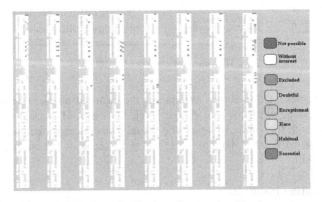

Fig. 12. Step pixelization of scene classification

Next, for each classification and for each scene, first, the absence of excluded valuations and the presence of indispensable valuation are tested as well as the fact of being complete.

Fusion of the 8 classifications enable, according to the parsimony paradigm - that says more a potential scene can solve the case, more it deserves to be favored -, to rank the potential scenes and to reevaluate the object classification (see figure 13) .

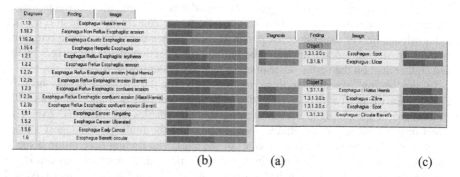

(b)   (a)   (c)

Fig. 13. Scene and object classification visualization: first the objects **(a)** next the scenes **(b)** and the object sorting reassessment **(c)**

Another challenge will be to traduce the classification and similarity retrieval operations directly in processing on the formal images issued of the pixelization of the databases.

4 Conclusion

A version of the endoscopic atlas is accessible via internet at http://i3se009d.univ-brest.fr/ - Password view2006-. At the moment, the application contains around 150 endoscopic images, 150 descriptions of diseases with 90 findings; it enables the user to describe an object or a scene in order to obtain the potential diagnoses and similar images. The response accuracy depends strongly on the user capacity to focus and analyze the endoscopic finding. Faced with an ambiguous diagnostic situation, it becomes difficult for the endoscopist to extract the relevant features from image. This inherent weakness is common to other computer-assisted decision-making systems and limits their relevance for diagnostic aid in real. On the other hand, the precise description of the cases by an expert, the flexibility of the research engine, the ability to store a large amount of cases and to use web facilities, as well as the improving knowledge of medical expertise aim the system towards a learning tool.

A study of the similarity in digestive endoscopy has been conducted next to endoscopists - novice and experienced – and next a witness group of students in psychology. The questionnaire contained a library of 90 well-known endoscopic images and 3 request images. For each request image, the participants were asked to choose the ten most similar images in the library and, for the endoscopists, to list the most relevant diagnosis. The comparison of the subject responses according to diagnostic accuracy suggests there are some relationships between the diagnostic accuracy, the perception of similar features, and the selection of images with the same expected diagnosis. One hypothesis could be that endoscopists choose images with similar diagnosis rather than in accordance with the aspect of the image. The choice of similar images would then depend predominantly on the diagnostic conviction of the endoscopist. The comparison of the subject responses according to their experience suggest that similarity between endoscopic images can be perceived, even by subjects who has no experience in that field of medicine. That suggests the hypothesis that similarity between images can be perceived, based on intrinsic properties of image content and without reference to knowledge and/or experience. A practical consequence could be that the retrieving method for an endoscopic atlas could be built on the selection of similar images by the user, rather than on the verbal description of the case under investigation.

In the setting of collaboration, another web site http://i3se009d.univ-brest.fr/ Collab - Email and Password: view2006 – is under construction and concerns 1500 endoscopic images. Already, the indexing module enables an identified expert to select images, to consult the clinical case, to describe the findings, to suggest a diagnosis or to select the most similar images of the atlas. The finding description is made easier by the matching to a similar image already indexed. Several experts can index the same picture and modify their description. All the discordances are displayed and the consensual validation only depends on the rules the team assert oneself. The experimentation module holds out the opportunities to develop some interactive tools and will enable to devise learning scenarios adapted to the reasoning way of the endocopist and to the experience level of the apprentice.

The endoscopy unit of the Brest Hospital purchased a Picture Archiving and Communications Systems (PACS). It will be coupled with novel endoscopic reporting software leaning on the Minimal Standard Terminology, still under development. A light version of the system with fewer features shall be integrated in order to index and retrieve images or sequences. Huge Case Bases will be constituted for the endoscopy and the colonoscopy, and for this last one, data-mining techniques associated with pixelization will allow creating a knowledge base.

The ongoing and in project works will all beneficiate of visualizing the data as images. Case pixelization with large case bases will enable a better apprehension of the visual and semantic similarity. A joint exploitation of the numerical and symbolic content of the endoscopic images is the goal to reach with new pixelization at stake.

References

1. Cauvin, J-M, Le Guillou, C, Solaiman, B, Robaszkiewicz, M, Le Beux, P, Roux, C: Computer-assisted diagnosis system in digestive endoscopy. IEEE Trans Inf Technol Biomed, Vol . 7. (2003); 256-262

2. Le Guillou, C, Cauvin, J-M, Solaiman, B, Robaszkiewicz, M,Roux,C: Digestive Endoscopic Scene Analyze. Proceedings of the IEEE-EMBS congress, Vol 4 (2001) 3855- 3858

3. Aamodt, A, Plaza, E. Case-based reasoning: foundational issues, methodological variations and system approaches. AI Communications Vol 7 (1994) 39-59

4. Crespi M, Delvaux, M, Schapiro, M, Venables, C, Zwiebel, F: Working party report of the Comittee for Minimal Standards of Terminology and Documentation in Digestive Endoscopy of the European Society of Gastrointestinal Endoscopy; Minimal standard terminology for a computerized endoscopic database. Am J Gastroenterol , Vol 91 (1996) 191-216.

5. Akdag, H, De Glas, M, Pacholczyck, D. "A Qualitative Theory of Uncertainty" Fundamenta Informaticae, Vol 17,(1992)

6. McCormick, B-H, DeFanti, T-A, Brown, M-D: Visualization in scientific computing. Report of the NSF Advisory Panel on Graphics, Image Processing and Workstations (1987)
7. Lawrence, S, Lee Giles, C, Bollacker,K : Digital Libraries and Autonomous Citation Indexing. IEEE Computer, Vol 32 (1999) (6) 67-71,
8. Spence,R: Information Visualization. Addison-Wesley (2001)
9. Card, S K, Mackinlay , J D, Shneiderman, B, Readings in Information Visualization: UsingVision to Think. Morgan Kaufman 1999
10. Börner, K, Chen, C,   Boyack, K: Visualizing KnowledgeDomains. Annu. Rev. of Information Science & Technology, (2003) 37
11. Barnes, S, Peck, A: Mapping the future of health care: GIS applications in Health care analysis. Geographic Information systems Vol 4 (1994) 31-34

Multiscale Scatterplot Matrix for Visual and Interactive Exploration of Metabonomic Data

Fabien Jourdan[1], Alain Paris[1], Pierre-Yves Koenig[2], and Guy Melançon[2]

[1] UMR1089 Xénobiotiques INRA-ENVT, Institut National de Recherche
Agronomique, France
Fabien.Jourdan@toulouse.inra.fr
[2] Laboratoire d'Informatique, de Robotique et de Micro-électronique de Montpellier,
LIRMM UMR CNRS 5506, France
{Pierre-Yves.Koenig, Guy.Melancon}@lirmm.fr

Abstract. We describe a method turning scatterplot matrix visualizations into malleable graphical objects facilitating interaction and selection of pixelized data elements. The method relies on density estimation techniques [1,2] applied through standard image processing. A 2D scatterplot is considered as an image and is then transformed into nested regions that can be easily selected. Based on Wattenberg and Fisher [3], and as confirmed by our experience, we believe users have a good intuition interpreting and interacting with these multiscale graphical objects. Bio-molecular data serves here as a case study for our methodology. The method was discussed and designed in collaboration with experts in metabonomics and has proven to be useful and complementary to classical statistical methods.

1 Introduction

Information visualization is a useful approach for the analysis and exploration of large multidimensional data sets. One major contribution of visualization when compared with other non visual approaches is to offer direct interaction with the data providing immediate feedback. Visualization can thus feed the analysis process and guide the user in the exploration or refinement of scientific hypotheses. This is of utmost importance in the study of biochemical data mostly because of the vast amount of experimental data that need to be sorted, but also because of its inherent complexity. The present paper focuses on a visualization technique helping the analysis of experimental data measuring the impact of specific molecules on metabolism. Visualization has been introduced in the processing chain involving biologists, statisticians and chemists and aims at bringing new insights on the studied phenomenon as well as on the experimental methodology itself.

In this setting, the raw data we need to study has a multidimensional character as it gathers numerous attributes describing both the phenomenon under study and the experiment itself. As we shall see, our methodology enters the pixel-oriented paradigm [4], as we suggest to lay the experimental data on a two-dimensional layout while offering the user the possibility to interact with

P.P. Lévy et al. (Eds.): VIEW 2006, LNCS 4370, pp. 202–215, 2007.

this 2D representation. We form images where pixels correspond to elementary data elements positioned according to numerical attributes. The real benefit of the interaction is achieved by allowing users to dynamically inquire about the underlying information. The 2D view is thus transformed into a more malleable material while preserving its statistical properties. The selection of subsets of experimental measurements is thus made easier while remaining faithful of the statistical phenomenon under study.

This work partly relies on techniques that were originally described and used for the exploration of relational data [5]. The similarity with our previous work is that we allow the user not only to explore and manipulate the processed scatterplot data, but also to directly act on this representation to inquire about the underlying data. In the present case, however, the experiments require that we build a representation gathering a series of scatterplots enabling the user to cross-examine different experimental conditions on the one hand, and observe the effect of drugs as time evolves on the other hand. Hence, the user actually views and interacts on a series of scatterplot matrix (see Fig. 7 for instance) all residing in a 3D data cube. Our approach is inspired from the seminal work by Becker and Cleveland [6] and shares similarities with that of Martin and Ward [7]. The idea of interacting with scatterplots through multiscale images was triggered by the work of Wattenberg and Fisher [3].

As we will explain, our visualization came as a mean of exploring and interacting with experimental data. The idea of combining simple image processing techniques together with scatterplot representations emerged from discussions with users where the need to easily select and highlight elements in numerous scatterplots became clear. The visual exploration and ability to interactively select elements from the scatterplots proved useful and complementary to classical statistical methods such as principal component analysis (PCA). Indeed, some phenomenons could only be observed by visual inspection and were left unnoticed by PCA and factorial analysis. It seems that the confidence assessed by classical statistical methods is offered at the price of leaving aside details that are nevertheless of interest when visually exploring the data.

2   Metabonomics

Metabonomics is usually defined as the study of changes in metabolite profiles as a result of a biological perturbation (such as disease or physiological stress)[1]. This is precisely what the experiment we will concerned with is about.

2.1   The Experimental Procedure

The experiment concerns mouse populations which are either normally fed, or go through a change from their usual diet. Another factor is also studied, where mice

---

[1] Compare with *metabolomics* which is the systematic study of the chemical fingerprints that specific cellular processes leave behind. That is metabolomics focuses on the metabolites found in a biological organism, which are the end products of its gene expression.

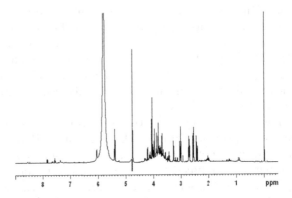

Fig. 1. The spectrum diagram describes the concentration (peaks) at which yet uniden-tified molecules appear in mice's biofluids

are administered a drug or a toxic chemical (a molecule or group of molecules), in addition to their diet. The challenge is to understand how their organism reacts to these perturbations (diet change and/or drug taking). The whole experiment follows a patented protocol [8] we shall briefly describe here.

The effect of the diet, with or without drug taking, is observed through sam-plings (blood or urine samplings). Mice are sampled a few times (once a month for two months and then more regularly, e.g.). Each sample is carefully pro-cessed and follows a series of steps to measure the presence of certain molecules (concentration) in reaction to the physiological stress the mice experience.

The samples are then analyzed to output a spectrum diagram measuring how much given molecules are present in the animal's organism (see Fig. 1) — one for each set of experimental conditions. At this step however, indirect measures only reveal the concentration at which some molecules are present in the organism. Extra work is needed to actually identify which molecules are present. Both steps have to be accomplished separately because concentrations are revealed through indirect measurements involving ion rays. Roughly speaking, the horizontal axis of the spectrum diagram (Fig. 1) is divided into sub-intervals according to the peak (concentration) structure. The largest peak on the left part of the diagram might for instance give rise to a series of sub-intervals. The ppm value associated with each sub-interval provides information that will help the identification of the associated molecule in a subsequent step. Sub-intervals are then considered as statistical variables that undergo a series of tests.

After collecting this numerical data, and after statistical analysis, the whole protocol leads back to the biological question, which is to identify pathways in-volved in the organism's reaction to physiological stress. Indeed, the presence and concentration of molecules in the samples relate to specific metabolic reac-tions that need to be discovered and studied. As the concentration of molecules vary from one situation to another, the biologist might infer hypotheses about how the organisms fights this stress (drug and/or diet change).

2.2   The Data and Task

The variables (related to molecule concentration) extracted from the spectrum diagram in Fig. 1 then undergo statistical tests (PCA) capable of identifying (with known statistical errors) groups of experimental conditions showing clearer contrast in terms of diet, drug taking and time elapsed after the experiment was initiated. A factorial analysis then allows the identification of the most significant variables with respect to the dominant PCA axis. In other words, the analysis is able to detect which molecules goes through the most notable variations under one or more experimental conditions.

Hence, the protocol produces a vast amount of figures. All mice populations (normal diet or unusual diet, with or without drug taking) are sampled six times, and approximately 750 molecules are traced in each of these situations. The complexity does not however come from this moderate data volume but from the need to simultaneously compare numerous pairs of situations. Indeed, because we are interested in studying molecules showing profiles that differ in distinct situations, we build scatterplots gathering profile information collected from two situations. Highly concentrated molecules in both situations are obviously of interest. The question however is to see whether this high profile of expression is observed in all situations, for instance, before enquiring about the pathways involved in the production of that molecule. Another scenario of interest is one where two molecules might systematically show opposite profiles (high concentration for one and low concentration for the other, or vice-versa, in all situations).

3   The Visualization

3.1   2D Scatterplots

The visualization we designed gathers 2D scatterplots mixing two sets of experimental conditions. More precisely, each axis spreads over the interval $[0,1]$ where $p \in [0, 1]$ corresponds to a normalized concentration (of a molecule in the organism). Hence, given a set of conditions S_1 (diet, drug taking and time elapsed) we assign each molecule m the concentration $p_{S_1}(m)$ at which it was observed. Now, given another set of conditions S_2 each molecule is mapped onto the 2D point $(p_{S_1}(m), p_{S_2}(m))$. Fig. 2 shows a typical 2D scatterplot. In this example, the two situations correspond to the same diet and drug taking conditions, but to different time intervals[2]. The x-axis correspond to molecule concentration after a month, while the y-axis correspond to molecule concentration after a bit more than three months. The position assigned to a molecule reflects its concentration in the two different experimental (set of) conditions. Two molecules can be mapped to the same 2D point. We take this into account by assigning points on a greyscale in order to reflect frequencies. This is of importance for what follows.

---

[2] From now on, we shall refer to the condition "time elapsed after the experiment was initiated" by "time intervals" or "time elapsed".

Fig. 2. Scatterplots are formed by crossing two sets of measurements (molecule concentration). In this case, the measurements concern the same mouse population, but were taken at two distinct moment in time. The x-axis corresponds to molecule concentration after one month, and the y-axis corresponds to molecule concentration after a bit more than three month.

It is in theory possible to build a scatterplot for all pairs of situations (set of conditions), leading to a set of (# diet types × # drug taking patterns × # of samplings) scatterplots like the one shown in Fig.2. However, we shall focus here on the evolution of molecule concentration over time. That is, we look at molecule concentrations for a given diet type and drug taking pattern (drug or no drug) after 32 days, 60 days, 88 days, and so on. We then form a (upper triangular) matrix where all scatterplots on row one map concentrations after day 32 on the horizontal axis, scatterplots on the second row map concentrations after day 60 on the horizontal axis, and so on. Conversely, scatterplots in column one map concentrations after day 32 on the vertical axis, etc. (See section 4.1.) We can form matrices for all possible pairs of diet type and drug taking patterns. Note that we choose to represent the diagonal. Indeed, it indicates how variables are distributed for a given condition. For instance on figure 10, selected molecules (red patches) spread differently through the 6 conditions).

Since axes are mapped to time, one could be tempted to look at the data using time series or parallel coordinates. However, time series or parallel coordinates did not offer a good readability of our data (which would typically require to be clustered) – see Fig. 3. Moreover, a time series visualization allows to easily read how the concentration of given molecules evolve throughout the whole experience, from day 1 to day 115 (mice are sampled 5 times). The interest of

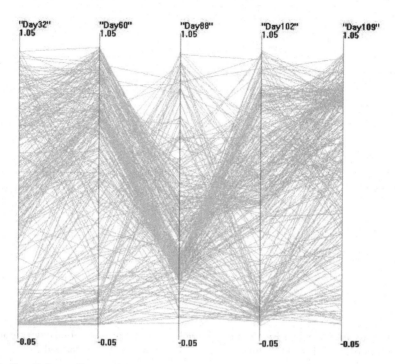

Fig. 3. Parallel coordinates (obtained using XMDVTool [9]) can be used to see molecule concentrations evolve along time

biologists however resides elsewhere. Their aim is not to model the variation of concentration of molecules, but rather to see whether the diet and drug pattern induce a clear change in the metabolism between two different samplings and, of course, understand why. Typically, these changes should appear as different spreading patterns of molecules on the scatterplots.

3.2 Gaussian Kernel and Blur

The fine-grained structure of the scatterplots makes them difficult to explore and manipulate. We borrow an idea from our previous work [5] initially inspired from Wattenberg and Fisher [3]. We consider the scatterplot as an image and blur it following a standard image processing technique. Each pixel, seen as a numerical value, is replaced by the result of a matrix convolution applied to its neighborhood. As the matrix encodes a gaussian kernel the pixel's neighbors contribute differently to the resulting value. This standard technique in image processing [10] actually performs an estimation of the density function encoded by the scatterplot [1,2]. The result of a gaussian convolution on the scatterplot of Fig. 2 is illustrated in Fig. 4[3]. Parameters of the gaussian kernel can be set to vary the blurred image.

---

[3] If we consider an image of NXN pixels and a kernel of size kXk, blur computation could be achieve in $O(K^2 N^2)$.

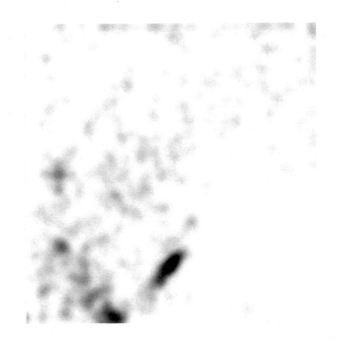

Fig. 4. The scatterplot in Fig. 2 is considered as an image and is blurred by applying a gaussian convolution. The blur is obtained from the scatterlot showed in Fig. 2.

3.3   Identifying Pre-selected Areas

The last step is to compute areas of relatively homogeneous greyscale in the blurred image. Molecules mapped to neighbor pixels in the scatterplot are merged into relatively homogeneous grey regions. This will help us identify and *select* regions gathering molecules behaving similarly. The regions are computed by performing image segmentation on the blurred scatterplot. This again is a standard image processing technique computing regions of neighbor pixels of similar greyscale value. The segmentation outputs regions delimited by closed curves, in a way similar to the level curves on a geographical map — we capture molecules having similar profiles, where similarity decreases as regions grow larger. The number of levels can be controlled in order to output larger or thinner regions. The right choice partially depends on the morphology of the original scatterplot.

Fig. 5 shows the resulting multilevel image after segmentation is computed on the blurred image (b) (all entries of the matrix in Fig. 8 also illustrate the result of image segmentation on various scatterplots). The final segmented image indeed reveals a multiscale structure as regions are all nested according to the different greyscales in the blurred image.

The segmentation is performed based on a partition of the greyscale range following ideas we shall now describe. First note that the greyscale values vary over the interval $[0, 256[$ (black pixels have a value of 0, and white pixels have a

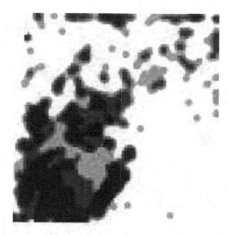

Fig. 5. Regions are computed from the blurred scatterplot by applying image segmentation techniques. The segmented image showed here is obtained from the blurred scatterplot in Fig. 4.

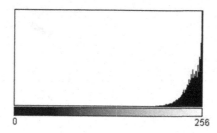

Fig. 6. Histogram describing how greyscale distributes over the pixels of the scatterplot in Fig. 2. The darker pixel has value 166. The histogram clearly indicates that a majority of pixels are pale grey.

maximum value of 255). Fig. 6 shows how the greyscale distributes in the blurred image (Fig. 4). Observe that the grey scale does not distribute uniformly over the full range $[0, 256[$, suggesting that a division into interval of equal length should be avoided.

Assume we want to segment the image into regions on k different levels. We want to find sub-intervals $[a_1, a_2[$, $[a_2, a_3[$, ..., $[a_{k-1}, a_k[$ (with $a_1 = 0$ and $a_k = 256$), such that the proportion of points lying in $[a_i, a_{i+1}[$ is $1/k$. This can be easily done using the inverse image of the density function (integral of the distribution in Fig. 6). When $k = 4$ this amount to finding what is often called the *quartile* of the greyscale distribution. This idea is commonplace in visualization and has been applied in other situations such as when defining a colormap over a set of data elements [11].

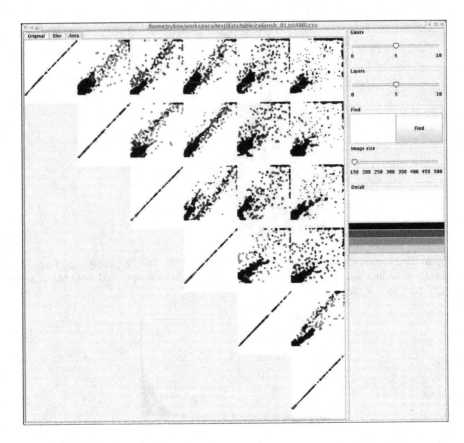

Fig. 7. Application snapshot showing the various interaction that can be performed. These interactions allow to tune blur parameters since suitable parametrization could evolve according to the dataset. The matrix shows (processed) 2D scatterplots of pairs of situations. Each line/column correspond to a different time interval elapsed after initiating the experiment (time intervals increase from left to right).

4   Case Study

4.1   Interaction

Fig. 7 shows a snapshot of our application. When starting the exploration, the user will set various experimental conditions defining the scatterplot matrix to be visualized. As mentioned earlier, we assume the user decides to let time intervals vary while keeping the other experimental conditions fixed (same diet/drug taking schema for all scatterplots). The visualization we built finds its full utility when the user is able to act on it and select various regions. To this end, the user may:

– vary the parameters of the gaussian kernel;
– vary the number of layers in the segmented image;

– click on a region and have access to the underlying data;
– activate regions containing any given molecule by performing a search.

Fig. 8 shows the scatterplot matrix gathering data collected from a population of mice that have *not* been administered *any drug*, but have only experienced a change from their usual diet. One thing is immediately observed when looking at the scatterplot images: those lying on the last two columns spread much further away from the diagonal than do the first columns. This observation agrees with the PCA analysis (see Fig. 9); using only PCA however requires additional work in order to determine the reasons underlying this cut. Following this observation, biologists were able to trace back methodological errors in mice diets. We see here one advantage when using the scatterplot matrix. Indeed, any statistical analysis of the data will be obscured by the diet event during the last two samplings. Using the scatterplot matrix however, users can simply decide to ignore the data from the last two columns.

Fig. 10 provides another example. In this case however, mice have all been administered a drug but have *not* experienced a change from their usual diet.

Notice the red patches that have been activated through all scatterplots. This activation has actually been triggered by the selection of a small red patch in the scatterplot sitting at position (4,5) on the fourth row and fifth column — the selected patch is pointed at by the arrow. This small red patch corresponds to molecules that are found at rather high concentration after the fourth time

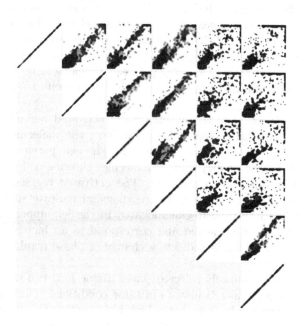

Fig. 8. Scatterplot matrix of mice experiencing a diet change. Observe that the data in the last two column spread further away from the diagonal.

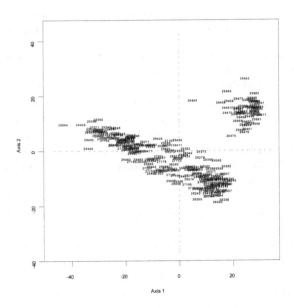

Fig. 9. PCA analysis clearly separates the variables into two groups indicating that mice diet experienced an unusual change

interval after the experiment has been initiated (high x-coordinate) but that show low concentration after the next time interval has elapsed (low y-coordinate). A search through all regions of all other scatterplots allow to automatically activate regions containing any molecule present in the selected patch. In a sense, the automatic activation of patches actually performs brushing [6,7]. The use of image blurring and segmentation however provides brushes specific to the dataset under study.

The visual inspection of the automatically activated regions is informative and can potentially bring knowledge about how the molecules relate to one another, or how they relate to metabolism. The red patch we selected (the one pointed at by the arrow) contain molecules showing a drop in concentration from a time interval to the next. The activated regions captures other types of variations. Regions close to the diagonal indicate similar concentration at both time intervals. Regions located in the left upper part show profiles dual to the one we selected and correspond to an increase in concentration. These variations might indicate a change of phase regulated by molecule concentration.

In other case, the initially selected patch divide into two distinct subset of molecules showing distinct behaviors in some conditions. This observation can not be inferred directly from statistical analysis, and on the contrary, is rather immediate when visually exploring the data. Note also that these observations seem more or less obvious because of their enhanced readability (as a result of the blur and segmentation processes).

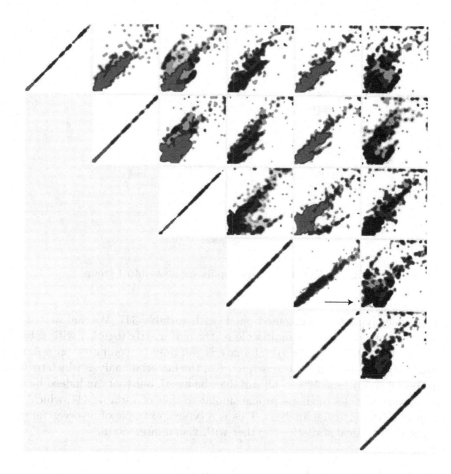

Fig. 10. A small red patch indicated by a pale blue oval has been selected in the scatterplot sitting on the first row and second column. The other red regions have been activated as they contain molecules underlying the selected patch.

4.2   Combining Classical Statistics with Visual Interaction

We shall provide evidence showing that visual exploration of the scatterplots comes as a complementary approach to classical statistical analysis.

PCA is used first to detect whether certain experimental conditions have a stronger and more determinant effect. Factorial analysis then splits variables according to these experimental conditions.

PCA underlined the fact that the most predominant factors explaining differences between molecule profiles were the last two samplings. The factorial analysis identified variable number 347 as being clearly involved in the metabolic changes. Using our tool we performed a search to locate variable 347 through all pairs of situations. As an output of the search, we also obtained the list of

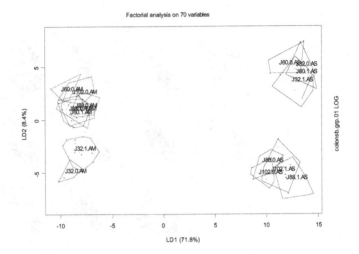

Fig. 11. Factorial analysis splits variables into 4 groups

all variables lying in a segmented patch with variable 347. We find here a second advantage of using our multiscale scatterplot matrix together with calssical analysis. Indeed, because factorial analysis can only be computed on a limited subsets of variables (this is a consequence of the factorial analysis which requires that between 30% and 50% of all variables be used), our tool can indeed be used to discover variables having a profile similar to that of variable 347, which were not part of the factorial analysis. This is a typical example of a successful combination of classical statistics together with visual interaction.

5   Conclusion and Future Work

We have described a method turning scatterplot matrix visualizations into malleable graphical objects facilitating interaction and selection of pixelized data elements. The method relies on density estimation techniques [1,2] applied through standard image processing. A 2D scatterplot is considered as an image and is then transformed into multiscale graphical objects where nested regions can be easily selected. Based on Wattenberg and Fisher [3], and as confirmed by our experience, we believe users have a good intuition interpreting and interacting with these multiscale graphical objects. It may well be that most people now have experience using geographical level maps.

This method has been introduced within an experimental protocol after experts have recognized its usefulness and complementarity to classical statistical methods. Our research did not focus on the design of new visualization techniques. Indeed, discussions with experts made it clear that we only had to assemble classical image processing techniques with scatterplot matrices together with basic interaction. An ongoing collaboration with all experts involved in the whole experiment (biologists, statisticians and chemists) will help the integration of this visual exploration method together with their existing practice[4].

Access to the underlying data is not yet completely satisfactory, mainly because of the inherent complexity of the experimental protocol. Also, more user feedback is needed in order to decide of various parameters that could be preset (radius of gaussian kernels, thickness of regions, etc.). Extension of our tool will include other types of kernels to let the final processed scatterplot vary. Again, user feedback should help measure the impact of these design choices on the usefulness and usability of the method.

We also plan to develop a view gathering all experimental conditions into a "datacube". The set of all scatterplots matrices obtained by letting all conditions vary indeed form a 3D or even 4D cube. The user might find useful playing with the cube in order to select particular experimental conditions when defining the initial scatterplot matrix.

References

1. Silverman, B.: Density Estimation for Statistics and Data Analysis. Chapman & Hall (1986)
2. Scott, D.W.: Multivariate Density Estimation : Theory, Practice, and Visualization. Wiley Series in Probability and Statistics. Wiley-Interscience (1992)
3. Wattenberg, M., Fisher, D.: A model of multi-scale perceptual organization in information graphics. In North, S.C., Munzner, T., eds.: IEEE Symposium on Information Visualization, IEEE Computer Society (2003)
4. Keim, D.A.: Designing pixel-oriented visualization techniques: theory and applications. IEEE Transactions on Visualization and Computer Graphics **6** (2000) 59–78
5. Chiricota, Y., Jourdan, F., Melançon, G.: Metric-based network exploration and multiscale scatterplot. In Ward, M., Munzner, T., eds.: IEEE International Symposium on Information Visualization, IEEE Computer Society (2004) 135–142
6. Becker, R.A., Cleveland, W.S.: Brushing scatterplots. Technometrics **29** (1987) 127–142
7. Martin, A.R., Ward, M.O.: High dimensional brushing for interactive exploration of multivariate data. In: IEEE Conference on Visualization '95, IEEE Computer Society (1995) 271–278
8. Dumas, M., Canlet, C., Debrauwer, L., Martin, P., Paris, A.: Selection of biomarkers by a multivariate statistical processing of composite metabonomic data sets using multiple factor analysis. Journal of Proteome Research **4** (2005) 1485–92
9. Ward, M.O., Rundensteiner, E.A., Yang, J., Doshi, P.R., Rosario, G.: Interactive poster: Xmdvtool: Interactive visual data exploration system for high-dimensional data sets. In: IEEE Symposium on Information Visualization. (2002) 52–53
10. Russ, J.C.: The Image Processing Handbook. 3rd edn. CRC Press (1998)
11. Herman, I., Marshall, M.S., Melançon, G.: Density functions for visual attributes and effective partitioning in graph visualization. In Roth, S.F., Keim, D.A., eds.: IEEE Symposium on Information Visualization, IEEE Computer Society (2000) 49–56

---

[4] To see other applications, please refer to this URL: http://www.lirmm.fr/~koenig/scatterplot.

ICD-View: A Technique and Tool to Make the Morbidity Transparent

László Daragó[1], Pierre P. Lévy[2], Anett Veres[3], and Zsolt Kristóf[3]

[1] Semmelweis University, Institute of Medical Informatics
Üllői út 78/B, 1082 Budapest, Hungary
darago.laszlo@gmail.com
[2] Departement de Sante Publique, Hopital Tenon, Assistance Publique Hopitaux de Paris,
INSERM U7071
4 rue de la Chine, 75970 Paris Cedex 20 France
Pierre.levy@tnn.aphp.fr
[3] University of Debrecen Health College Faculty, Health Information Department,
Sóstói út 2, 4400 Nyíregyháza, Hungary

Abstract. Morbidity and mortality are one of the most frequently used of statistics. It is hard to oversee the whole distribution of cases because the diagnoses spread on too many International Classification of Diseases codes (ICD 10th version). Usually, in practice, hospital managers are satisfied to study some pre-determined and/or ordered groups of data. ICDview will help to find these groups. The distribution of diagnoses is varying by time, location and several parameters. The ICD codes themselves have a main structure by the organ and kind of disease. Examining the cases by this technique can also show, which ICD classes must be regrouped.

Caseview method bases on the Diagnosis Related Groups (DRGs). Some-hundred pixels are pictured. The number of elements of the ICD tops more than ten-thousand. Because of this it is problematic to pixelize it, because it is very hard to picture such a big amount of data on one screen.

ICDview uses the same like reference set, as the Caseview does. The ICD main groups are classified in the Caseview's columns. The groups of these main groups are pictured in these columns, but their order from the midline separating the medical and surgical entities can be determined on several ways. One way is to calculate it by the average prognosticated weight number of the cases at DRG. It can be calculated by the average prognosticated length of stay of the patients in hospital, and other ways too. The cases can be represented by the main diagnosis, the basic diagnosis or any other type of diagnosis, which stays in the background of the current cure.

Examples of use of the method are given.

1 Introduction

The hospital cases and inpatient cases are the basis of any mortality and morbidity statistic. There exist a general and overall data collecting system in those health care systems, which apply Diagnosis Related Group (DRG). The data, collected via the

P.P. Lévy et al. (Eds.): VIEW 2006, LNCS 4370, pp. 216–224, 2007.

finance system, contains a multitude of information about the health state of the population, the functioning of the health care both on nosological and economical levels. When a large amount of data is set to be evaluated, usually some hypotheses are examined. A key to have good evaluations and proof is to have good hypothesis.

The Caseview technique [1] is very useful to get ideas on what set of cases to examine that will support the hypothesis. It pixelizes the cases by their DRG. There are local and Internet applications installed to study the DRG itself, or a determined set of cases. France, Hungary and Belgium already created their own Caseview's. French Caseview was presented in MIE2003, St. Malo, 2003[2], Hungarian Caseview on PRO-ACCESS: 2nd International conference in e-health in common Europe, Krakow, 2004[3], while the Belgian Caseview was presented onPCS/E 20th International Working Conference, Budapest, 2004[4].

Another application of this technique is ICPCview[1]. It was presented in MIE2005, Geneva, 2005[5].

The Caseview is not for comparing international data sets. It is based on national code system. There is no reason to create an international Caseview application until the MBDS's[2] are unified.

The Caseview and ICPCview do not represent the mortality and morbidity rates, because they may be "economically motivated" at coding[6].

However, we need morbidity and mortality statistics. We also need international comparisons on the level of diagnoses. The case's main diagnosis, which determines its DRG, is unable to give a real nosological picture, however the basic illness of the patient can. We also have diagnostic data from several sources so the claim to pixelize the diagnoses arose as a real need.

2   Materials and Methods

In the planning and developing process of the ICDview, it should be kept to the forefront, that these techniques assume and support each other in analyzing and evaluating the morbidity and mortality rates. Because the users are the same, the principal ideas of the pixelization and the user surface must be compatible.

2.1   Planning the ICDview: The Idea and the Reference Set

The number of the diagnoses are too high to picture on one screen, thus a multi-level grouping was needed. The ICD-10 itself is organized into main groups, groups, subgroups and finally, the ICD classes. In this structure there exists 21 main groups which contain 240 groups, 2030 subgroups are contained in the 240 groups, and 10795 ICD classes are within the subgroups. The meaningless classes are ignored[3].

The number of groups can be viewed on one screen, so the resolution on first approximation is about 1:50. This way, the ICDview shows one ICD group as one pixel, the fine structure of the group can be showed by clicking the group. When the

---

[1] International Classification for Primary Care.
[2] Minimum Basic Data Set: the data, which represent a hospital care on national level.
[3] Sine morbo, Sine diagnosis, etc.

group's subgroups appear you may click on any subgroup and its fine structure and the classes will be shown.

As previously written, the entire *view should be oriented to the same structure, especially, to the Caseview's structure[4].

The Caseview's construction contains 3 main rules:

1. group the columns by nosological aspects
2. divide the columns into medical and surgical sides
3. locate the DRG's position within it's side by the weight number[5].

The ICD's main groups are sorted to the Caseview's columns. (This is the simplest way, but the principle must be overviewed in the future.) Doesmedical/surgical property of an ICD class exist? It is a question of preference. ICDview can separate medical and surgical cases. Another question is how to determine the ICD's order from the midline. A DRG-ICD cross-reference can answer on the last two question. A given ICD class belongs to several DRGs, while many ICD classes determine (besides other properties) one DRG. This is a more-more context. The DRG also has a property denoting whether it is medical or surgical. Making a cross-reference, that is, which ICD class can appear in which DRG's properties and showing the difference between the medical and surgical DRGs, the average weight number and, what can be ordered to the given ICD class, can be calculated. That means that an ICD can appear both on the medical and the surgical side.

This is one way to make the reference set of the ICDView. Other ways can be use the weight numbers compared to the number of cases in a period. The average length of stay of the patients in hospital can be used as any other nominated or weighted property of the DRG. The goal of the application determines the creation of the reference set within the Caseview-correspond rule. (Relating to the DRG origin provides that the ICDview also has some national dependency, while the national DRG has built in. International applications must be independent from local properties.)

Another aspect of the planning of the ICDview came from the Information System planning theory. It is critical that the user surface of the family of applications should look identical. It is very important because the ICDview is both a research tool and a teaching tool too. All these aspects determine a user surface which is very similar to Caseview_HUN.

Steps of developing the ICDview
1. Copy the ICD10 handbook to a spreadsheet, edit and structure it by the main group-group-subgroup-class level decomposition.
2. Import the spreadsheet into a database table.
3. Copy DRG classification handbook to word processor file.
4. Convert DRG classification rules from document to spreadsheet.
5. Import DRG classification rules spreadsheet to a database table.
6. Order medical/surgical status to ICD levels.
7. Order ICD main groups to Caseview columns.

---

[4] Caseview-correspond rule.
[5] The estimated cost/income, ordered to the DRG case.

8. Modify ambulant ICD class to medical class.
9. Eliminate pre-main groups[6].
10. Make the cross reference: 8104 of 10795 ICD classes can appear as main diagnosis at DRG classification. 169061 element in DRG-ICD cross reference.
11. Calculate medical and surgical sides and "weigh numbers", to sort the ICD classes in the columns.
12. Make the reference set: ICDV_hunicd.txt. It is be used every time, when open the application.
13. Make the Internet application, build a Caseview similar user surface.

As Fig.1. shows, the outer back colors of the pixels are adjustable by setting the weigh number limits, while the inner back color are adjustable by setting the inner limits. The pixels contain the evaluated numbers, such as number of cases, etc. The default value is the ICDs "weigh number", that is the reference set, itself. The decimal places can be set too.

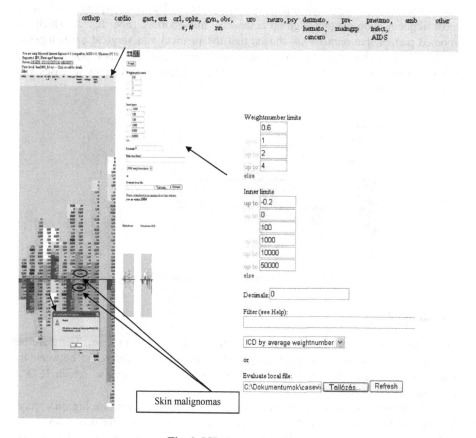

Fig. 1. ICDview user surface

---

[6] Pre-main groups can belong to several main groups, not allowed here .

The description of the ICD class appears by clicking a pixel.

It is possible to upload and evaluate the user's own file – this the real power of this application. The use of the WEB application is absolutely free and anonymous. The ICDview application is available at

http://mail.de-efk.hu/~darago/icdview/

URL, and other URLs, linked from that website.

3   Results

Some results demonstrate the usability of ICDview, in the following.

3.1   Result 1

Fig. 1. shows the Hungarian MBDS of all cases in 2002 (a base year, with approximately 3 million departmental inpatient cases). The first surprise was that the most frequent main diagnosis group is the skin malignomas. (See fig. 2., which is a zoomed part of that area.) This shows, that the medical and surgical cases together totaled 137099 departmental inpatient cases. The Caseview does not show this extreme value because the DRG classification procedure spread these cases into several DRGs and columns.

Fig. 2. Zoomed part with skin malignoma groups

3.2   Result 2

The main diagnosis and the basic illness of the patient are also shown. There are big differences within the same ICD groups. This defined the difference between the main diagnosis and the basic illness' diagnosis. This can preview, which ICD to evaluate.

Fig 4. Illustrates that differences also can be evaluated. There are big differences between the basic illness and the main diagnosis of the cases at some diagnoses.

Behind some of these big differences hide pseudo problems but some of them worthy of deeper analysis. The basic diagnosis and the main diagnosis have a high case number equality in obstetrics cases (O).

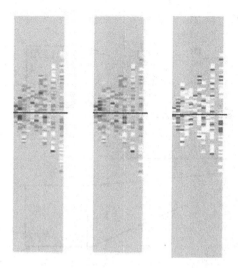

Fig. 3. Global view of 2002 cases: main diagnosis, basic illness, difference. Global view is the "mini-me" picture of ICDview. On the left is the view of the main diagnosis of the cases in 2002, the middle is the basic illness of the patient in the case, while the right figure is the difference between them. The colors of the cells are by the inner limits in order: deep green, light green, white, light yellow, yellow, magenta. The areas in the squares are zoomed in the next figure. Fig. 4. contains the marked areas on the main field.

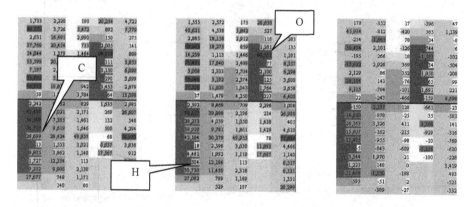

Fig. 4. Zoomed view of 2002 cases: main diagnosis, basic illness, difference

Hypertension (H) occurs much more frequently as the basic illness, while cerebrovascular or cardiovascular diagnoses (C) occurs much more frequently as main diagnosis of the care – this fact confirms, that how dangerous the "silent killer" remains.

"H" marks the hypertension, "C" the cerebrovascular and cardiovascular diagnoses. "O" marks the obstetrics cases.

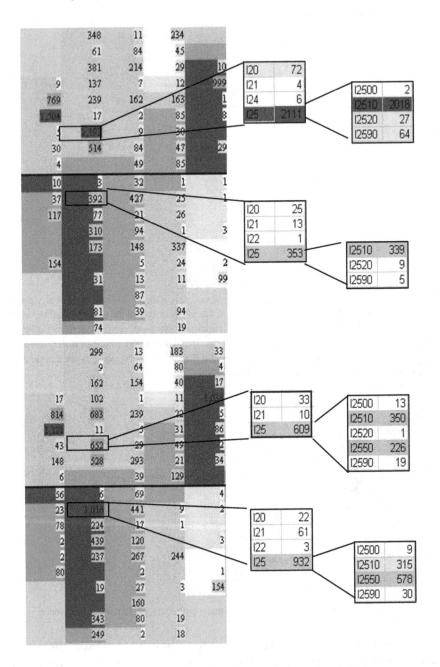

Fig. 5. Hospital one and hospital two, with fine adjust of the ICDview

3.3 Result 3

This result gives an example, how to use the ICDview tool.

Let's compare the cases of two hospitals. The cases are representing a full year. The hospitals are in the same economical and geographical area. It can be supposed, that the patients are about the same with their illnesses. The medical doctors graduated from the same universities and the medicines and medical treatments also the same. Both of the hospitals are supported by the state and the financial system is the same. Perfect subject to make an "internal analogy".

Hospital one had twenty-one-thousands cases while hospital two had twenty-three-thousands. Both of them work with about five hundred beds.

The zoomed part of the hospitals' ICDview are figured on Fig. 5.

In the first view, the distribution of the diagnoses are the same, no big differences appear. Except, the marked places, where at hospital one, the number of patients in the I20-25 group (Ischemia hearth diseases) is much bigger in the medical part than the surgical. Hospital two has an opposite state, that is, the surgical part is more predominant than the medical. Why such a radical difference between the two hospitals? Why does hospital one have so much more hearth disease compared to hospital two and to the national level?

The next table summarizes the hospitals' interest

Table 1. Hospital one, hospital two and the national data

	Yearly cases	Medical I20-25	Surgical I20-25	Rate of medical I20-25/all cases	Rate of surgical I20-25/all cases	Rate of medical/ surgical I20-25 cases
Hospital one	21531	2193	392	10,18%	1,82%	17,88%
Hospital two	23732	652	1018	2,74%	4,29%	156,14%
Country wide	2818801	93946	78417	3,33%	2,78%	83,47%

3.3.1 Different Treatment

The reason that hospital one had less surgical cases is because a medical university is located next door and they transfer their patients for intervention. The big number of medical cases may be increased as the patients come back after the intervention. In this way the before intervention and after intervention cases appear twice. One extra before or after them. (That can be a problem for the insurance company as what to and what not to pay.)

The state of hospital two is more interesting. The surgical cases weight number is very high. This fact can arose the suspicion that this large declination from the national level proportion is not occasional.

3.3.2 Different Diagnostics

There may be differences in the type of care and there may be differences in the diagnoses. Fig. 5. also shows that the results of the diagnoses differ as well Hospital one does not have patient with I2550 code, while this ICD is the most frequent code in hospital two.

This article does not want to decide any of questions above. The may results show a need for an overall review in these hospitals.

4 Conclusion

There is a vast amount of information included but hidden in the data of health care. ICDview pixelizes and pictures the diagnostic data, and the user may recognize some set of data, which can add ideas to the making of hypotheses. This paper briefly described the idea and the way to use the tool, to represent the idea.

Acknowledgement

Here we thank to Dr. István Bordás for his help to make the cross-reference set. We are very grateful to Victoria Miller for her efforts at checking and correcting the English of this paper.

References

[1] Lévy PP. Le case view une methode de visualisation du case mix. Journal d'économie médicale mars 2002; 20(2), 118-127.

[2] Lévy PP: The case view a generic method of visualisation of the case mix. International journal of Medical Informatics 2004, 73, 713-718.

[3] László Daragó: Caseview_HUN: easy DRG overview. Studies in health Technology and Informatics. Volume 105. Transformation of Healthcare with information technologies. IOS Press. Editors: Marius Duplaga, Krzysztof Zielinski, David Ingram.pp182-189, 2004

[4] Pierre Lévy, László Daragó, Claire Beguin, Francis H. Roger France: application of the Caseview method to the Hungarian, Belgian and French DRG classification, PCS/E 20th International Working Conference, 2004. October 27.-30.

[5] Pierre P. Levy, Laetitia Duché, Laszlo Darago, Yves Dorleans, Laurent Toubiana, Jean-Francois Vibert, Antoine Flahault: ICPCview: Visualising the International Classification of Primary Care, Connecting Medical Informatics and Bio-Informatics, Proceedings of MIE2005, Geneva, Switzerland, IOS Press Nieuwe Hemweg 6b 1013 BG Amsterdam , 623-628., 2005.

[6] Erik Sverrbo, Ola Kindseth: Economic motivated coding of medical data. The effects of activity based financing and strategies to prevent an unwanted development in Norwegian somatic hospitals, PCS/E 20th International Working Conference, 2004. October 27.-30.

[7] http://www.b3e.jussieu.fr/caseview/drgview

[8] http://www.medinfo.hu/darago/caseview.php

[9] Lévy PP. Caseview : building the reference set. In: Studies in health Technology and Informatics. Volume 105. Transformation of Healthcare with information technologies. IOS Press. Editors: Marius Duplaga, Krzysztof Zielinski, David Ingram.pp172-181, 2004.

[10] László Daragó, Pierre Lévy: Caseview_hun: un outil internet pour appréhender les DRG hongrois, Journal d'Économie Médicale, 2003. Vol 21. no 7-8, 451-454.

[11] http://www.medinfo.hu/darago/icdview.php

Pixelization and Cognition

Time Frequency Representation for Complex Analysis of the Multidimensionality Problem of Cognitive Task

Montaine Bernard[1], Noël Richard[1], and Joël Paquereau[2]

[1] Laboratoire Signal-Image-Communications, Bât. SP2MI,
Bd Marie et Pierre Curie, B.P. 30179, 86962 Futuroscope Cedex, France
{bernard,richard}@sic.univ-poitiers.fr
[2] Equipe Pulsar, Pôle Biologie Santé, 86000 Poitiers, France
joel.paquereau@chu-poitiers.fr

Abstract. Brain functioning comprehension is an actual challenge. However the cerebral recordings produce a huge amount of data, it is difficult to select the pertinent data with the studied task. As we favor the dialog with the expert, we developed a methodology based on visualization. All steps are discussed and validated by the neurophysiologist as they still keep the link with the biological process.

The presented method is based first on extraction of oscillatory information from time-frequency transform. These ones are organized in a graph structure. Finally graph-matching techniques bring indication signals variations such as frequency, time or power variations.

1 Introduction

As the Mars planet exploratory expedition, the brain is one of the most important investigations of the next decade. Nevertheless the cerebral activity complexity, because of a huge data set, has a multidimensional formulation with 3D localization, various frequencies and time variations. These electromagnetic activations, essential to brain functioning comprehension, directly reflect information transmission among neurons. By EEG, it is possible to observe these very short activations (duration of only some milliseconds) with a high spatial resolution, up to 256 electrodes on the scalp.

The actual developments on brain analysis use classical signal processing approach to describe or model low level activity. However these are not able to produce a high level description of brain activations. The main difficulty in such research is to link mathematical results to expert knowledge. Thus, in our work, we have given greater importance to simple graphical representation at each process stage to compare our results and the understanding of the neurophysiologist.

The EEG signals embody the brain complexity. To analyze them it is necessary to have regard for the cerebral information coding way along all the dimensions. Cerebral processes are coded by amplitude, time, frequency and spatial variations which describe the different complexity level of a complete task (from stimulus

P.P. Lévy et al. (Eds.): VIEW 2006, LNCS 4370, pp. 227–239, 2007.
© Springer-Verlag Berlin Heidelberg 2007

to reaction). Moreover intra and inter-individual cerebral variations should be considered.

First, we present in this article the principle of information transfer analysis and the classical techniques to study this topic. Then we describe the new approach to analyze brain functioning. Finally a discussion will be led on advantages and disadvantages of this new method in comparison with the classical ones.

2   Classical Brain Information Transfer Analysis

The brain can be regarded as a highly parallel and distributed information processing system. This implies the existence of distinct neural pathways leading to different locations, working side by side, extensively interconnected and never converging on a single common area. Two levels of information integration are present [1]:

local integration: neurons populations are tuned as they own specific properties. By the mean of EEG, we observe **synchronization** as these neurons populations overweight the rest of not-synchronized neurons [2].

large scale integration: constant interaction with each functional areas. This phenomena is studied by analyzing the **synchrony** among the different neurons populations responses from these areas located at different spatial positions [3].

A task in which the subject had to hear a question, simultaneously see an information then think about the answer and finally answer is a good example to conceive these two levels of integration. In this process, some local areas are activated in a definite frequency band to do a specific task (the primary auditory cortex is activated to hear the sentence for instance). These activations are due to the neurons synchronization. And then to answer the subject had to integrate the verbal and visual information. This integration is done by a neural synchrony phenomena. This oscillatory phenomena (in gray on the EEG signals of the figure 1) is not easily delimited. The figure 1 shows what we would like to display: the activated areas and the relationships between them.

As EEG is a complex signal with both stochastic (non-stationary) and deterministic (stationary) properties, estimate the information transfer between cortical areas is not an easy task. Many different methods are available to analyze this neural populations synchrony. These methods can be roughly divided into two basic categories: parametric such as auto-regressive models assuming that EEG is generated by a specific model and non-parametric methods such as conventional spectral analysis assuming no specific model for EEG generation. In this part, we present the main techniques of information transfer analysis according to the interpretation they provide whatever the methodology they are issued.

2.1   Cross-Correlation

This is one of the oldest and most classical measures of interdependence between two signals. The cross-correlation function measures the linear correlation

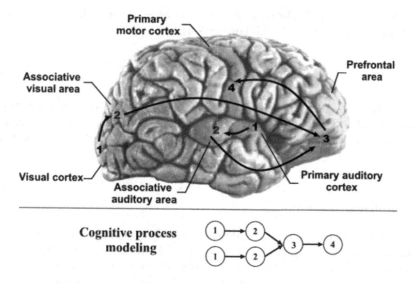

Fig. 1. Example of brain processes display

between two signals $x(t)$ and $y(t)$ as a function of delay time (τ), which is of interest because such a time delay may reflect a causal relationship between the signals. However this is not necessary always the case, since internal delay loops of one of the systems or different distances to the source may change this interpretation [4].

$$C_{xy}(\tau) = \frac{1}{N - \tau} \sum_{k=1}^{N-\tau} x(k + \tau)y(k) \qquad (1)$$

where N is the total number of samples.

The energy function $C_{xy}(\tau)$ takes a maximum when the two signals are correlated at the position of the delay τ due to the information transfer.

However, the accuracy of cross-correlation is strongly dependent on the length of analyzed segment and on the level of additive noise. With EEG signals this length is most of the time too short and too much noise is present.

2.2   Coherence and Phase Analysis

To avoid limits correlation about noise and signal length, the coherence calculation is based on techniques using model paradigm.

Mathematically, coherence is calculated from signal power spectrum which is typically done by Fourier transform [5] or auto-regressive model [6] or wavelets transform [7]. In the equation 2, S denotes the spectral estimate of two EEG signals x and y for a given frequency f. The numerator contains the cross-spectrum for x and y (S_{xy}) ; the denominator contains the respective autospectra for x (S_{xx}) and y (S_{yy}). For each frequency, coherence values lie within a range from 0 (no correlation) to 1 (full correlation).

$$Coh_{xy}(f) = \frac{|S_{xy}(f)|^2}{S_{xx}(f)S_{yy}(f)} \tag{2}$$

While the coherence spectrum provides the strength of correlation between the two signals $x(t)$ and $y(t)$, time delay information between two signals can be obtained from the phase spectrum, that is the argument of the cross-spectrum S_{ij} [8]:

$$\phi_t(f) = arg\{S_{xy,t}(f)\} \tag{3}$$

2.3   Multivariate Analysis

The methods discussed so far are defined for only two signals: a functional relationship obtained by paired-wise analysis of bi-variate signals. However for practical neural data analysis the activities are recorded from multiple spatial positions in the brain, so one can create a multivariate modeling framework containing all available information from different channels. Let $X(t) = (X_1(t), \ldots, X_D(t))'$ denote a multivariate time series from D data channels. The multivariate auto-regressive model describes the signal in each channel as a linear combination of its own past activity and the past activity of all other channels plus an additional uncorrelated noise. Consequently the data epoch properties are described by the fitted MVAR model parameters A_k [9].

Direct transfer function. The directed transfer function (DTF) is a normalized version of the transfer function $H(f) = [\sum_{k=1}^{p} A_k exp(-2\pi i k f \delta t)]^{-1}$ with p the model order, f the given frequency and δt the sampling interval. [10]. It describes the ratio of the influence of component X_j on component X_i to all the influences on component $X_{l\neq i}$. Due to the normalization, the DTF takes values in $[0, 1]$.

$$DTF_{ij}(f) = \frac{|H_{ij}(f)|^2}{\sum_{m=1}^{D} |H_{im}(f)|^2} \tag{4}$$

Partial directed coherence. Using a different normalization than Kaminski et al., Baccala et al. defined the PDC [11] with $\bar{A}^{-1}(f) = (I - A(f))^{-1} = H(f)$:

$$PDC_{ij}(f) = \frac{|\bar{A}_{ij}(f)|}{\sqrt{\sum_{m=1}^{D} |\bar{A}_{mj}(f)|^2}} \tag{5}$$

2.4   Interpretation

To explain how to interpret the values coming out these coupling functions, we show the results on simulated data. The simulation is composed of five electrodes defined by the five equations specifying the link among each electrode. The simulated links are presented on the figure 2.

The electrodes couple maps show the existent link between the two concerned electrodes, with the origin electrode in column and the destination electrode in row. The link power is on the y-axis and the frequency band $(0-60Hz)$ on the x-axis. The resulting schemes present that the coherence values show some additional cooperation (green) whereas the DTF shows indirect and direct links and PDC only direct links (Fig. 2).

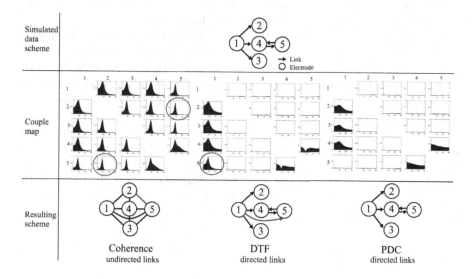

Fig. 2. Summary of links found among the five simulated electrodes

These presented methods define the relationships among cortical areas. All of these approaches are based on a statistical activity modeling for each axis. Since the model parameters are not understandable, it is difficult to construct a discussion with the specialist. Hence we propose an alternate established on the ability to maintain the model parameters understanding.

3   EEG Signal as Graph Structure

The main idea of the representation is that the brain response to stimulation is not a linear response but a sum of linear responses in a non linear structure. The hypothesis is that the low level activity producing synchronizations could be model by linear response. A non linear structure is used to link these synphronizations. The non linear aspect is embedded in a graph and the countinuous aspect in a time-frequency analysis.

3.1 Signal Modeling

In order to obtain a relationship between the low level activity and the high level processing, a good EEG signal description is required. Since the size of the recruited neurons populations acts on the EEG signal amplitude and the task acts on the frequency band of information coding, we chose a time-frequency representation with a hierarchical analysis based on energy level.

Time Frequency representation. The wavelets are mathematical functions that cut up data into different frequency components, and then study each component with a resolution which is adapted to its scale [12].

The chosen *mother wavelet* $\psi(t)$ is the complex Morlet wavelet which has a simple analytic form and has a good time and frequency resolution [13]. Moreover this wavelet has an easy interpretation and thus contributes to facilitate the discussion with the neuroscientist. The complex Morlet function is the product of a complex exponential wave and a Gaussian envelope where constant k is the wavenumber defined in accordance with the signal length (see eq. 6).

A *wavelet family* $\{\psi_{a,b}\}_{a,b\in\mathbb{R}^{+*},a\neq0}$ is the set of elementary functions generated by dilatations and translations of the mother wavelet $\psi(t)$ where a, b are the scale and translation parameters, respectively, and t is the time. The *continuous wavelets transform* of the signal $s(t)$ of finite energy is the family of the coefficients $C_f(a,b)$ defined as a correlation between the function $s(t)$ and the family wavelet $\psi_{a,b}$ for each a and b, ($\overline{\psi}$ is the complex conjugate of ψ):

$$C_f(a,b) = \int_{-\infty}^{+\infty} s(t)\overline{\psi_{a,b}(t)}dt, \qquad \psi(t) = \pi^{-\frac{1}{4}}e^{ikt}e^{-\frac{t^2}{2}} \qquad (6)$$

The wavelet estimate of the energy density in time-frequency plane -*scalogram*- is given by $E^{a,b} = |C_f(a,b)|^2$. By this translation, we represent our EEG signal in a time-frequency map as shown on the figure 3 (b); the high energy level is represented by a red color and the low one by a blue color. Consequently, if a red burst appears on the map, it means that this frequency is strongly presents at this given time (for instance, lots of 10 Hz at 0.2 sec).

Conserved information. The Morlet discrete continuous wavelet produces nearly a continuous image closed to the continuous digitized signal. The neurophysiologist is interested in some typical signal patterns which are represented as energy bursts in our time-frequency map. An adaptive threshold is applied on the time-frequency map to remove the noise and background activity. This threshold is defined for each map according to the map histogram . This histogram is bimodal, one mode, in the low level, pixel represents the noise and the other mode, in the high level pixel, represents the event-related activity. Consequently a bi-Gaussian modeling is used to define the threshold (Fig. 3 (c)). The removed pixels (lower than the threshold) can be considered interfering noise or neuronal background without link with the studied task. After all, a watershed algorithm [14] is use to segment the thresholded map into bursts of interest

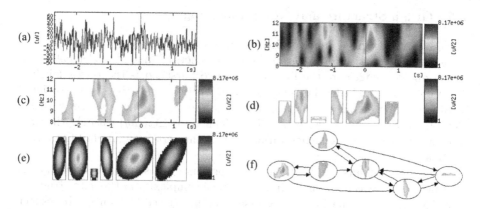

Fig. 3. EEG signal modeling. (a) the raw EEG signal, (b) the time-frequency map from a Morlet wavelet decomposition 8-12Hz, (c) the adapted thresholded time-frequency map, (d) the bursts extracted from the thresholded time-frequency map by a watershed algorithm, (e) the gaussian modeling the extracted bursts, (f) the graph built from the extracted bursts.

(Fig. 3 (d)). This time-frequency map reduction has been validated by comparison by Event-Related Sycnhronisation [15].

Signal features set. To sort the information, segmentation is applied on the time-frequency map. The signal features are specified first from extracted bursts of the segmented map and second from bursts gaussian modeling (Fig. 3(e)). By this way, it is possible to build a multi-scale approach of brain activity analysis intelligible for the neuroscientist. If B is the bursts set and $|B|$ its cardinal, each segmented burst i $(i = 1, \ldots, |B|)$ is individually characterized by (x_i, y_i) the time and frequency position of the energy maximum of the burst, $(\hat{\mathcal{E}}_i, \bar{\mathcal{E}}_i)$ the maximum and averaged energy of the burst, θ_i the angle between the horizontal and the burst main direction and (σ_i^x, σ_i^y) the standard deviation along the ellipse major and minor axis.

To conclude an EEG signal is summed up by a set of bursts feature F. The different steps of EEG features are presented in the figure 3.

$$F = \bigcup_{i=1}^{|B|} \left(x_i, y_i, \hat{\mathcal{E}}_i, \bar{\mathcal{E}}_i, \theta_i, \sigma_i^x, \sigma_i^y \right) \tag{7}$$

Even if the amplitude and frequency information has been investigated by the burst extraction, no information is kept about their relationship and spatial position. We chose to use a graph structure able to describe on the one hand the energy burst localization in time, frequency and spatial space and, on the another hand the dependence relation, i.e. the activated areas and the links between them. Graphs are a way to keep a structural continuity of bursts in all dimensions.

3.2 Graph Structure and Graph-Matching

Graph definition. A labeled directed graph is a quad $G = (V, E, \alpha, \beta)$. The elements of V, vertices of the graph G are the energy bursts, the elements of E, graph directed edges are the link between two bursts. α is the vertex labeling function and β the edge labeling function. Vertices and edges are associated with labels that describe their properties. L_V and L_E are the labels set with L_V defined by $(\hat{\mathcal{E}}_i, \bar{\mathcal{E}}_i)$ of one vertex i and L_E defined by $(\Delta x, \Delta y)$ with Δx the time variation and Δy the frequency variation from the vertex origin until the vertex destination.

To chain these bursts inside the graph representation is our main difficulty. As a connection between all bursts implies a huge complexity in the data structure and the graph manipulation, it is necessary to limit the edges number. To respect the brain activity and enhance the structure ability, we only consider edges in which the time distance between the two bursts is low in accordance to the detail level (frequency domain). Consequently the structure allows us to manipulate 2-complex representation (Fig. 3(f)).

Graphs comparison. Comparing two cortical activities corresponds, in our case, to compare the variations of the recorded EEG answer in term of latency, frequency, energy and activated areas. Thus, estimate these variations, with our data structure, is equivalent to measure the graph similarity. For that two steps are necessary, the first one is to find the most common subgraph in the two graph candidates, it is the graph matching problem. The second is the measure computation between these two common subgraphs. Generally a unique optimization algorithm is used to combine these two steps, the graph matching problem using a similarity measure.

These similarity functions have to be defined to know how similar two bursts or edges are. For each label $l \in \{L_V, L_E\}$, a similarity function $S_l \in [0, 1]$ is defined. The actual similarity formulation use L1 expression. The purpose was to validate the structure and the approach, not the distance functions or similarity formulation. Moreover numerous existing formulations are possible from Minkowski Norm to fuzzy expressions. Consequently a specific work on this question is in process.

Then it is possible to estimate the similarity $S_{G_1 G_2}$ between two graphs G_1 and G_2. The goal is to maximize this similarity to find the most similar common subgraph.

$$S_{G_1 G_2} = \sum_{l \in L_V} \sum_{v_1 \in V_1} \sum_{v_2 \in V_2} M(v_1, v_2) S_l(v_1, v_2) + \sum_{l \in L_E} \sum_{e_1 \in E_1} \sum_{e_2 \in E_2} M(e_1, e_2) S_l(e_1, e_2)$$

$$(8)$$

with M the result matrix of the best matching between G1 and G2. The implemented matching algorithm is based on the A* algorithm [16] which permits to find the optimal solution. However it is not optimized in term of computation time ($O(\log(n))$ complexity, with n the number of graph vertices). Some algorithms as those proposed by Gold or Ranganath [17,18] could be more optimized.

Time delay estimation. The graph matching gives a similarity measure between two graphs i.e. two signals. It can be view as a measure of coherence between the two signals, it tells us how similar, how coherent, are the two signals. To give an example, we apply this measurement on real data.

Ten subjects were confronted sequentially with the stimulus material. In detail, the temporal sequence of trials is:

1. a question of approximation,
2. the thinking to elaborate a solution to the raised question,
3. the answer to the question ,
4. a pause of 5 seconds minimum.

The neurophysiological knowledge brought some information about the brain processes involved in this sequence. Three main tasks should be found, first the question hearing activating the auditory cortex then the thought activating the prefrontal lobe and last the answer enunciation activating the motor cortex.

The objective of the experimentation was to measure the activation delay between each area of interest. We chose the middle electrode of each area: P3 in the auditory cortex, F4 in the prefrontal lobe and Cz in the motor cortex. We chose also a neutral electrode F7 to compare the results. Then we matched the burst structure issued of these four electrodes. The following results are one example of what could be obtained. The similarity measure is higher between the first three electrodes P3, F4 and Cz than with F7.

By studying the time shift between each graphs couple with a high similarity measurement, we found a delay of 0.9s from P3 to F4 and of 3.8s from P3 to Cz (Fig. 4). So the time delay is going from P3 to F4 and later from P3 to Cz. We found again the same process as the one described by the neurophysiologist. And the results appear to be in the same range as the physiological expectations.

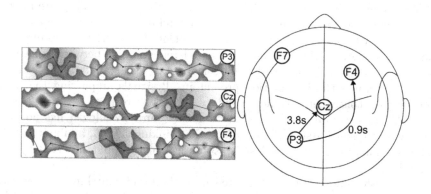

Fig. 4. Time delay expected by graph-matching is conformable with processes described by neurophysiologist. The studied tasks are a question hearing activating the auditory cortex then a thought activating the prefrontal lobe and last an answer enunciation activating the motor cortex.

4   Discussion

Two points are important in the brain functioning imaging, first the brain activated area description and secondly the brain information transfer description.

By way of the signal modeling, cerebral activated areas are observed. As a matter of fact the used time-frequency transform allows us to visualize oscillatory phenomena. The extraction of these ones sums up the signal by a set of features characterizing the neural synchronization. This synchronization is synonymous of brain activation. The bursts synchronization examination gives the active brain areas. To validate our approach, we have compared our results with results from a well-known technique. The event-related phenomena represents frequency specific changes of the ongoing EEG activity and may consist in general terms, either of decrease or of increase of power in a given frequency bands. This may be considered to be due to a decrease or an increase in synchrony of the underlying neuronal populations, respectively. The former case is called event-related de-synchronization (ERD) and the latter event-related synchronization (ERS). A method has been developed, mainly by Pfurtscheller et al. [19], to measure these phenomena. A statistical wilcoxon parametric test was used to know if the ERD/ERS results found from the burst structure is significantly different or not of the ERD/ERD from method of Pfurtscheller et al. The test was performed on the 1000 original signals (40 trials on 25 electrodes). The wilcoxon results show that the two methods give results not significantly different (with $\alpha = 0.01$ and $p = 0.4395$). Moreover one important advantage of the signal modeling use is the huge data reduction. For instance, the size of a compressed file,with a duration of 10 minutes, 128 electrodes and a sample frequency of 1024Hz, is about 300Mo whereas the modeling resulting file is about 70Ko.

To understand the brain functioning, it is important to be able to analyze the relationship between the different implicated areas. The first section of this article describes the actual methods to analyze this problem. They are mainly based on statistical approach. That implies a good technique to describe a large set of data with no a a priori on the treatment. However, two assumptions had to be done first the necessity of a good repeatability between each trial and a high number of trials to have a certain statistical validity. This kind of methods averages the information present in all signals amplifying the main process; as a consequence some secondary processes are lost. For example, in the real data presented in the last section, the neurophysiologist has noticed, visually, a frequent burst in the middle of the thinking (Fig. 5, in about in 25% of trials). This information is lost by statistical method but modeled by our method.

The transfer information is deducted from the graph-matching results. But a deeper analysis had to be performing to study the relationship described by this analysis. At present we are not able to know if we describe only the direct links between the activated areas or also the indirect links (with intermediate). Moreover the graph-matching analysis is not limited to the time shift analysis (i.e. information transfer analysis). In fact all the features will be compared; it

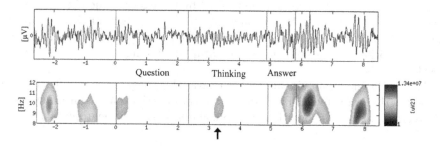

Fig. 5. Frequent burst in the middle of thinking (8-12Hz, Cz). By classical analysis this pattern is lost by the averaging process. However the graph-matching could be a way to study it and analyze its function.

means that an analysis on the frequency and energy variation is in progress. Such an analysis brings information about brain rhythms variations and the size of the mobilized neurons populations, respectively. On the other hand, we need comparative data to check and validate our method. We had to find for the same large number of acquisitions the same main processes with an equivalent relationship description.

Our main preoccupation about this work is to favor the dialog with the expert. To reach this goal we promote all the means to simplify the communication between the two communities. The best way stays a good display. For this reason why we try to visualize our partial results at each step of the computation and discuss about their validity with the expert (Fig. 3). Concretely we control all process analysis as we know at each step how the data are used (for instance the burst comparison in the graph-matching). Moreover by way of this visualization, we can also analyze single-trial which is not possible with statistical methods.

5   Conclusion

The results obtained with the presented method are encouraging. However some improvements had to be brought, mainly in the graph structuring and graph matching. First of all, the graph edge building is still a problem. The different building possibilities had to be analyzed in a deeper way and tested with simulated data. Secondly the similarity functions used in the graph matching process are actually very simple. If they allow a structure validation they do not have an acceptable behavior. Our current work is based on a theoretical expression of the distance function on the time, frequency and spatial domain. As the physiological mechanisms are different for the information transfer (spatial aspect), transmission and reaction time (time aspect) or rhythm variations (frequency aspect), a specific formulation is needed for all these dimensions. Lastly some optimized matching algorithm can improve the computation speed.

The present work is to develop a method describing the brain functioning with a possible visualization at each step of the processing. Consequently each result at any time can be checked and interpreted by the expert. One difficulty in our method is the graph-matching complexity, some knowledge of the neurophysiological domain is necessary to reduce this complexity. As statistical methods need no a priori, we had to compare our graph-matching results with statistical results.

At present, no brain signals model is available, only a set of rules can try to define the brain functioning. Image and pixelisation paradigm bring a link between high levels of cognition concepts and low levels of neurophysiology.

References

1. Varela, F., Lachaux, J.P., Rodriguez, E., Martinerie, J.: The brainweb: phase synchronization and large-scale integration. Nat Rev Neurosci **2**(4) (2001) 229–39
2. Elul, R.: The genesis of the EEG. Int Rev Neurobiol **15** (1971) 227–272
3. Friston, K., Stephan, K., Frackowiak, R.: Transient phase-locking and dynamic correlations: Are they the same thing? Human Brain Mapping **5** (1997) 48–57
4. Pereda, E., Quiroga, R.Q., Bhattacharya, J.: Nonlinear multivariate analysis of neurophysiological signals. Prog Neurobiol **77**(1-2) (2005) 1–37
5. Rappelsberger, P., Petsche, H.: Probability mapping: power and coherence analyses of cognitive processes. Brain Topogr **1**(1) (1988) 46–54
6. Nunez, P.L., Srinivasan, R., Westdorp, A.F., Wijesinghe, R.S., Tucker, D.M., Silberstein, R.B., Cadusch, P.J.: EEG coherency. I: Statistics, reference electrode, volume conduction, Laplacians, cortical imaging, and interpretation at multiple scales. Electroencephalogr Clin Neurophysiol **103**(5) (1997) 499–515
7. Lachaux, J.P., Lutz, A., Rudrauf, D., Cosmelli, D., Quyen, M.L.V., Martinerie, J., Varela, F.: Estimating the time-course of coherence between single-trial brain signals: an introduction to wavelet coherence. Neurophysiol Clin **32**(3) (2002) 157–74
8. Schack, B., Rappelsberger, P., Vath, N., Weiss, S., Mller, E., Griessbach, G., Witte, H.: EEG frequency and phase coupling during human information processing. Methods Inf Med **40**(2) (2001) 106–111
9. Marple, S.: Digital Spectral Analysis with Applications. Prenrice Hall, Upper Saddle River, New Jersey (1987)
10. Kaminski, M., Blinowska, K., Szclenberger, W.: Topographic analysis of coherence and propagation of EEG activity during sleep and wakefulness. Electroencephalogr Clin Neurophysiol **102**(3) (1997) 216–227
11. Baccala, L.A., Sameshima, K.: Partial directed coherence: a new concept in neural structure determination. Biol Cybern **84**(6) (2001) 463–474
12. Durka, P.J.: From wavelets to adaptive approximations: time-frequency parametrization of EEG. Biomed Eng Online **2**(1) (2003) 1
13. Meyer, Y.: Wavelets and operators. Volume 37 of Cambridge Studies in Advanced Mathematics. Cambridge University Press, Cambridge (1992) Translated from the 1990 French original by D. H. Salinger.
14. Beucher, S., Meyer, F.: The morphological approach to segmentation: The watershed transformation. In: Mathematical Morphology in Signal Processing. Marcel Dekker Inc., New York (1993) 433–481
15. Bernard, M., Richard, N., Paquereau, J.: Functionnal brain imaging by eeg graphmatching. (2005) 27th annual conference of the IEEE Engineering in Medecine and Biology Society(EMBC'05), Shanghaï, Chine.

16. Nilsson, N.: Principles of Artificial Intelligence. Symbolic Computation. Springer (1982) NIL n 82:1 1.Ex.
17. Gold, S., Rangarajan, A.: A graduated assignment algorithm for graph matching. IEEE Trans. Pattern Anal. Mach. Intell. **18**(4) (1996) 377–388
18. Ranganath, H.S., Chipman, L.J.: Fuzzy relaxation approach for inexact scene matching. Image Vision Comput. **10**(9) (1992) 631–640
19. Pfurtscheller, G., da Silva, F.H.L.: Event-related eeg/meg synchronization and desynchronization: basic principles. Clinical Neurophysiology **110**(11) (1999) 1842–1857

Instant Pattern Filtering and Discrimination in a Multilayer Network with Gaussian Distribution of the Connections

Dimitri M. Abramov[1,2] and Renan W.F. Vitral[1,3,*]

[1] Center for Computational Intelligence, Adaptive Systems and Neurophysiology - NIPAN,
Department of Physiology, Biological Sciences Institute, Federal University of Juiz de Fora,
Brazil, 36.036-330
[2] Laboratory of Cognitive Physiology, Institute of Biophysics, UFRJ, Brazil
[3] ICONE-LSI, Department of Electronic Systems, Electric Engineering and Polytechnic
School, USP, Brazil
renanvitral@ieee.org
http://www.renanvitral.org

Abstract. This paper was designed to build an artificial multilayer network with the purpose of studying abilities like instant pattern recognition and discrimination where no learning would be required. The relevance refers to: (1) theories about putative biological mechanisms that would support innate perception, (2) technological implementation of faster systems for detection and classification of environmental stimulus without learning. Our model was built using few paradigmatic principles of neural organization. The connections obey a Gaussian function. When the network is submitted to diverse input patterns it produces both discriminative and distributed codes in all layers. Contrasting stimulus leads to an attention-like process by salience detection. Finally, the codes always hold a half of all nodes.

1 Introduction

The computational abilities to stimulus discrimination in either biological or artificial intelligent systems have been largely associated to learning [1, 2, and 3]. This way, several computational models need encrypt memory through learning algorithms to develop pattern classification. In biological systems feature detection is clearly improved by learning strategies [4].

Here we raise the following questions: is learning required to perception? Is perception *per se* an evolving ability of the nervous system? Is perception acquired or would be innate?

The creation of artificial assemblies capable to categorize stimuli patterns without learning would be important as tools for: (1) theories about putative biological paradigms that would support innate perception, (2) technological implementation of faster systems able to detection and classification of environmental stimulus without learning.

---

* Corresponding author.

P.P. Lévy et al. (Eds.): VIEW 2006, LNCS 4370, pp. 240–252, 2007.

An emergent discriminative code related to some stimulus would be built by a feed-forward system able to organize a distributed representation well-suited to some input features. Perception (selected by an attention process) is represented in the neural systems probably through a distributed coding [8, 20] (i.e., synchronization of neural units in a coherent collective functional state [9]). Authors have advocated to a sparse coding approach to distributed representations [20]. The combination of the activity of the synchronous units in a distributed coding has been elected as the best strategy to information representation [6, 7]. Furthermore, the segregation of image features to be processed by different cortical areas (e.g. shapes are coded through occiptotemporal stream while movement is detected into occiptoparietal cortex) supports the prediction of a distributed coding to visual information, which would be consciously perceived as a whole. Theories have proposed combinatorial strategies of the dynamical neuronal activity to code taste on the time, when similar stimuli would be able to produce similar activity patterns [10, 11, and 15]. The smell representation seems to be distributed on several processing levels in the olfactory system, probably as a spatial map, and this coding schema seems to be similar for both, invertebrate and vertebrate olfactory systems [12, 13, and 20]. Discriminative and distributed representations seem to be generated on olfactory bulb in mammals [14]. Authors have shown sparse coding in the visual areas [16, 18] and rat auditory system [19].

On primate brain, the visual fields fall in higher visual areas, where there are broader distributions of activity and modality-specific cells [17].

In this theoretical approach, we seek build an assembly capable to organize emergent distributed coding, without learning algorithms and informative about stimulus features so that these ones could be identified even when presenting minor distinctions, as we shall see on pixelized images. We looked for study how stimuli combination would be represented on the output regarding stimulus features as its contextual contrast or eccentricity on the input.

2   The Model

First, we do not wish to regard our network units as neurons but as processing units on a network array capable to share excitatory or inhibitory connections among them. Beyond, primary visual and sensory pathways brain regions seems to be architecturally organized in columns [28].

It were built non-plastic connections in a four-layered processing network with 900 units (30x30 units) in each level, which establish a range of three millions of couplings at all (see diagram in figure 1 to a representation of this network). Our model works in a simple way: the first layer receives the stimuli and drives only one-way connections to the second level (layer); the units inside in each one of the layers 2, 3 and 4 connect among them through lateral excitatory and inhibitory couplings, and each unit drives connections to the units in the immediate next level, such as input layer does. Biological evidences suggest that sensory systems establish a functional hierarchy [21], so we modeled our inter-level connections such as to emulate a hierarchical system. We assume that the strength of lateral and inter-level

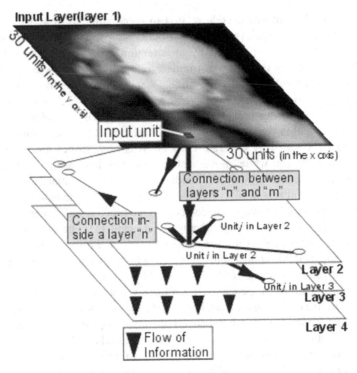

Fig. 1. Schematic view of the model

connections tends to be reduced while the spatial distance between units extends. Observing optimization of neural connectivity and wiring economy, models based on biological evidences regard a low probability of connections between distant nodes (which can be cortical regions or another nervous structure or even neurons), yielding clustering [24, 26, 27]. The coupling probability between units in a small neural network could decrease obeying a Gaussian function related to the distance [22, 23, and 27]. Here we used this probabilistic function as a rule for the establishment of the connections.

The network units work through non-linearity between activity thresholds like in the BSB (brain-state-in-a-box) model [29]. The unit answer must be regarded as proportional to neuronal spike frequency. We must notice that our answer function only drafts a paradigmatic neuronal behavior. It is obvious that cells generate spikes under diverse and particular conditions. To emulate it, it would be required simulation of diverse kinds of neuronal populations.

The connection C, between two units i and j located inside layers n and m, is defined by:

$$C_{ijnm} = k \times S_{ijnm} \times N_{ijnm} \times A_{ijnm} \qquad (1)$$

where $k = 0.1$ (arbitrary), S (fixed as 1 or -1) sets if a coupling is inhibitory (-1) or excitatory (+1) ($n = m$ if two distinct units are in the same layer). Biologically, we consider the nature of effective connection between two units (i.e., the net effect that a unit set upon another): an excitatory neuron that excites an inhibitory interneuron that connects to another unit has an effective inhibitory role upon that unit. A is another random value, between 0 and 5 (also arbitrary), and N sets the tendency to dispersion of connections, pointed as:

$$N_{ij} = 1 \qquad\qquad\qquad\qquad if \quad i_x = i_y,\ j_x = i_x,\ n = m \qquad (2)$$

$$N_{ij} = \frac{2.5}{(2.\pi s)^{\frac{y_2}{2}}} e^{-\frac{[(i_x - i_y)^2 + (j_x - j_y)^2 + d]^2}{2.s^2}} \qquad\qquad otherwhise$$

where x and y ($x = 1, ..., 30;\ y = 1,...,30$) are the Cartesian coordinates to an unit inside the 30x30 array of the layers. We set $s = 9$ and $d = 1$. The value to parameter s yield a Gaussian curve with specific length and high. The parameter d is the distance between i and j inside the z axis in a 3-dimentional Cartesian space. By this equation, greater the spatial distance between i e j, weaker the connection.
It was fixed a minimum threshold to the connections, where:

$$\begin{aligned}C_{ijnm} &= 0 \qquad if \qquad C_{ijnm} < 0.01 \text{ and } n = m \\ C_{ijnm} &= 0 \qquad if \qquad C_{ijnm} < 0.02 \text{ and } n \quad m\end{aligned} \qquad (3)$$

These conditions reduce the couplings between layers. Studies suggest clustering on small-word networks, where large-connected assemblies share fewer connections among then [24, 25].

First layer: input layer. Here the units do not share lateral connections. The input pattern is printed as the initial state X of each unit i that yield answers as follows:

$$Y_i = X_i \qquad (4)$$

Each unit i in the layer n receives inputs from j units in the preceding layer $(n - 1)$ to produce a state labeled as $X'_{i(n)}$, as follows:

$$X'_{i(n)} = \sum_{i=j=1}^{N = 900} Y_{j(n-1)} \times C_{i(n)j(n-1)} \qquad (5)$$

Next step: the initial activity X' for i sets answers Y' by statement (7). And the following algorithm is processed to yield a new state X, which is computed, based on these states for each i. So, the lateral couplings are processed as follows to yield a new state:

$$X_{i(n)} = \sum_{i=j=1}^{N = 900} Y'_{j(n)} \times C_{i(n)j(n)} \qquad (6)$$

Y is the input value from j to another unit i. The following conditions determine Y (or Y') of a unit based on its X (or X'):

$$\begin{aligned}Y_i &= 0 \qquad if \qquad X_i \le 0 \\ Y_i &= 3, \qquad if \qquad X_i \ge 3 \\ Y_i &= X_i, \qquad\qquad otherwise\end{aligned} \qquad (7)$$

2.1 Input Patterns and Output Analysis

Each input pattern was a 30x30 array of given states X to respective unit i in the input layer ($0.00 \leq X_i \leq 5.10$). Inside a same input pattern we may define distinct stimuli. The stimuli are unit assemblies whose states are $X > 0$. We used input patterns with one or more stimuli inside them. The current input patterns are in the figure 2 and 3: they varied on eccentricity, size, intensity, conformation or internal contrast (*i.e.*, the pattern of stimuli is defined by the difference of contrast among their units); and also multiple stimuli, which were different to size, intensity and eccentricity. Observing the input pattern at figure 02-G, one would regard two stimuli inside it. However, there is no spatial disjointing between these supposed different stimuli (they are contiguous) to denote two different objects. The visual system is able to segregate

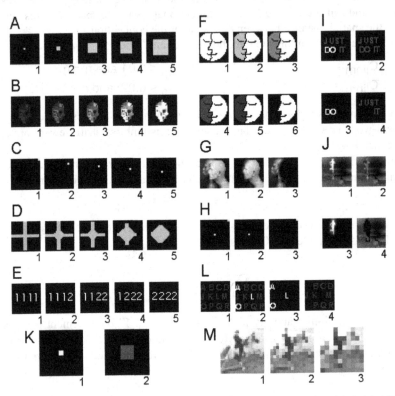

Fig. 2. Input patterns to the model (256 shades of gray). Group A: squares with 4, 16, 100, 144, 256 units respectively (same intensity = 4, 1); group B: varying intensity; C: stimulus (4 units) displacement to periphery; D: pure form transformation; E: spatial transformation of a composite stimulus; F: contrast-based form transformation; G, H, I: composite pattern with displaced or intense stimulus to study coding concurrence (credits to R. Mapplethorpe for fig. G-1); J: contrasting stimulus to a uniform background to study emergent stimulus segregation, L: disperse intense word among another weaker ones to study behavior to non-local stimulus; K: stimulus with different size but same total energy; M: same picture with different resolutions (respectively, 30, 15 and 10 dpi).

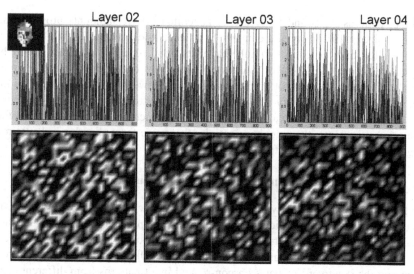

Fig. 3. Outputs from each processing layer. Top: histogram showing noisy distribution of unitary activities throughout the 900 units. The total activity slows down throughout the layers. Down: the respective activity maps (white = max. activity; black = no activity).

stimulus features to interpret it as two different objects [33]. The same process is important to discriminate an object from its context [34]. The figure 02-M contains stimuli with three different resolutions: 30, 15 and 10 dots per inch (dpi).

Once it was established the input patterns, our interest was to study the activity pattern on each layer produced by these input patterns. Thus, we had made the following parameter analysis: output size (counting all units i that showed $Y > 0$), output total activity (is the sum of all Yi values in a respective output pattern), output map (it is a comparative study of the activity maps produced by input patterns, which sets similarity indexes between two output patterns, hamming distances). The Similarity Index (SI) is *1 minus* the binary difference between two output patterns, divided by *I*, where each active unit was set as *1*. The following prescription sets SI:

$$SI = 1 / (1 - \text{hamming distance}) \tag{8}$$

3 Results

3.1 Output Patterns – Aspect, Size and Activity

As expected, the network creates distributed output patterns that had a noisy aspect for any layers by any input pattern. The activity was uniformly spread on the layers. Any layer produced outputs with nearly the same size for any input pattern, turning on 48% of all units (mean): the lower value that we found was of 409 active units, and the greater one was 456 active units. The output length varied less than 6%. Inside this short gap, there are no relation between stimulus intensity, size or location and the output length. As expected, the output maps had a noisy aspect for any layers by any input pattern. The activity was uniformly spread on the layers.

Input patterns A, B and C (figure 2) generated activity amounts proportional to stimulus size and intensity and contrary to its eccentricity (table 1). We can observe that smaller and eccentric stimuli have its total energy increased throughout successive layers (e.g. C-2) or it does not alter, not occurring to large or intense stimuli. Eccentricity does not produce differences between stimuli closer to the input center (note that C-3 activity is bigger than C-4, and these both are higher than the A-1 activity). Despite the total intensity (the sum of activity of all input units) for K inputs to be the same, the amount of activity that arose from stimulus K-1 was much bigger than total energy at output by K-2 stimulus.

3.2 Output Patterning – Similarities Between Output Maps

We have compared two-by-two the output maps produced by each input pattern from all groups. Once there is a wide range of possible comparisons, it will only be shown the congruence among outputs produced by patterns from a same group. Initially, we had discovered the minimum stable state, the smallest similarity index through averaging the congruencies between non-related input patterns from different groups (e. g. comparing similarity between output maps produced by input patterns D-1 and E-5). Using t-test, we compared the mean values for layer 2 (0.695 ± 0.07, mean $\pm$ SD, n = 661 comparisons) to layer 3 (0.680 ± 0.06, $p < 0.001$) and to layer 4 (0.679 ± 0.06, $p < 0.001$). However, we can find very small SI between input patterns for the same group at first layer: SI – First layer (A1 x A5) = 0.58; SI – Second layer (A1 x A5) = 0.63 (see table 2 to all computed SI).

The similarity indexes between inputs from group A, where only stimulus size varies, reflect input likeness and this behavior is more prominent at layer 2, even with smallest SI values for A1xA5 and A2xA5. The SI values at fourth layer are all

Table 1. Activity by input patterns A, B, C and K

Input Pattern*	Activity. Layer 2	Activity. Layer 3	Activity. Layer 4
A-1 (04 units)	150	147	168
A-2 (16 units)	313	338	317
A-4 (144 units)	672	565	495
A-5 (256 units)	806	617	525
B-1 (942)	784	622	525
B-2 (667)	683	581	481
B-3 (423)	507	488	429
B-4 (280)	347	367	349
C-1 (16 units)	73	87	89
C-2 (11 units)	111	128	145
C-3 (06 units)	152	161	179
C-4 (03 units)	164	192	215
K-1 (09 units, 27)	176	186	202
K-2 (81 units, 27)	55	64	63

(*) The number of the units refers to the stimulus size (for A, at units), intensity (B) and eccentricity (C). Both, size and intensity are shown to K.

larger than those at others levels. Input patterns from group B, which have the same form and size but their intensity have varied, produces SI values also larger for closer intensities in the input patterns. We can notice that weaker input patterns yield closer maps. The SI values by group B are larger than those by group A, and unlike this group, inputs that varied intensity alone do not generate different SI values along the layers. To group C, the eccentricity seem to produce different SI alone at first layer, this behavior was not so obvious.

Table 2. SIs between input patterns – groups A to F

	Ly 02	Ly 03	Ly 04		Ly 02	Ly 03	Ly 04
A1 x A5	0.58	0.66	0.71	D1 x D2	0.91	0.90	0.90
A2 x A5	0.63	0.69	0.71	D1 x D3	0.97	0.86	0.87
A3 x A5	0.71	0.74	0.78	D1 x D4	0.76	0.80	0.81
A4 x A5	0.77	0.78	0.80	D1 x D5	0.72	0.74	0.77
B1 x B2	0.96	0.96	0.96	D4 x D5	0.83	0.81	0.83
B1 x B3	0.92	0.92	0.90	E1 x E2	0.83	0.82	0.82
B1 x B5	0.83	0.80	0.82	E1 x E3	0.74	0.75	0.79
B2 x B5	0.83	0.81	0.82	E1 x E4	0.71	0.74	0.79
B3 x B5	0.85	0.83	0.84	E1 x E5	0.66	0.70	0.75
B4 x B5	0.92	0.89	0.88	E4 x E5	0.84	0.84	0.87
C1 x C2	0.62	0.68	0.74	F1 x F2	0.95	0.94	0.94
C1 x C3	0.57	0.65	0.71	F1 x F3	0.91	0.89	0.89
C1 x C4	0.57	0.65	0.73	F1 x F4	0.86	0.86	0.85
C1 x C5	0.58	0.65	0.74	F1 x F5	0.80	0.81	0.82

This network produces output SI values proportional to input similarity. For group D (a progressive transformation of a form) this behavior is evident at all levels; however discrimination is better at last layer. On the contrary, observing behavior by group E, we can note that discrimination occurs at first layer. For group F the SI differences are evident at all layers and reflect stimulus disparities.

In the table 03, the SI values are listed to input patterns at the groups G to J in the figure 01. The group G we are comparing a stimulus to its parts and it is notable that more intense and large part and the whole stimulus have a larger SI. Similar comparison is made at the group J: the small and contrasting part of input pattern is separated from its broad and soft context so the whole input is more similar to its contrasting part than to its context. When the contrast is lowered in the input pattern J2, this pattern is more similar to stimulus background. The inputs from group I and L bind different stimuli in the same pattern and there is a similar behavior to group J: the more contrasting and smaller stimulus produced bigger SI when compared to whole input pattern than to input pattern with bigger ant less contrasting stimulus. In the group L, the input 2 is more similar to input 3 than to input 1. To this group the behavior is more prominent.

We also compared input patterns with an identical stimulus at different eccentricities. There is a notable SI to stimulus at center and input with a central plus eccentric stimulus, which is not observed to other combinations.

Finally, we observed that reducing resolution of a 30 dpi picture to 15 or 10 dpi, the SI was larger than 30 dpi picture to control ones (see figure 2 group M). The system was more efficient to recognize the 15 dpi picture as congruent to original stimulus.

Table 3. SI values between input patterns – groups G to J

	Ly 02	Ly 03	Ly 04		Ly 02	Ly 03	Ly 04
G1 x G2	0.90	0.86	0.87	H1 x H2	0.90	0.86	0.89
G1 x G3	0.65	0.70	0.78	H1 x H3	0.63	0.73	0.76
G2 x G3	0.56	0.64	0.74	H2 x H3	0.55	0.65	0.72
J1 x J3	0.75	0.75	0.83	L1 x L2	0.79	0.79	0.81
J1 x J4	0.78	0.77	0.78	L1 x L3	0.68	0.71	0.75
J2 x J4	0.84	0.82	0.81	L1 x L4	0.87	0.85	0.88
J3 x J4	0.56	0.63	0.68	L2 x L3	0.88	0.88	0.89
I1 x I2	0.85	0.83	0.84	L3 x L4	0.58	0.64	0.71
I1 x I3	0.84	0.86	0.88	M1 x M2	0.93	0.93	0.92
I1 x I4	0.69	0.71	0.76	M1 x M3	0.90	0.88	0.88
I2 x I4	0.82	0.83	0.86	M2 x M3	0.91	0.89	0.89

4 Discussion

Although the output patterns show activity to a half of all units into layers it does not seem to be organized as clusters. The binary index to the output maps varied nearly in a range of 20%. Thus, a minimum of 70% (with exception) of the units does not change their activity state, and the network computes stimulus discrimination using till 30% of all units on any layer: This model effectively works with sparse codes [7]. The first layer uses significantly more units to code the input patterns. Nearly 70% of units are commonly active for any output (see, for example comparisons D1xD5, E1xE5, G2xG3). This system seems to establish a SI minimum at nearly 0.7. So, we regard that this model produces sparse codes.

This model discriminates between stimuli regarding its similarities or its displacement inside the input pattern organizing distributed codes although with related similarity degrees. To stimuli displacement, the network behaves as follows: the second layer is able to discriminate relative distances between locations better to bigger stimuli than smaller ones; the output layer realizes suitable discrimination to different locations to bigger stimulus alone. The SI values are globally lower to activities at the second layer. The SI values also are lower to small stimuli. These observations suggest that the early levels of this system are more sensible to discriminate stimulus location than higher ones and all system is more sensible to smaller stimuli, which suffers a larger relative displacement. Our observation suggests that the higher level is able only to discriminate the stimulus nature (as its form or intensity). Three different degrees of resolution are tested by this network (see group M). The ability for generalization of this network could be a powerful tool to reduce the cost (memory load, for example) during image processing to target detection. If a lesser pixel concentration (resolution) is needed, the system is optimized. The other key points are: (1) not-needing of the pattern learning; (2) higher velocity of processing (since occurs within only one interaction of the net algorithms).

Different stimuli placed together as input patterns seem to compete to be coded. Besides, the more contrasting stimulus leads to more activity to output pattern and drives the input encoding. In fact, this network is very sensible to contrast (i.e. background intensity versus stimulus intensity). K input patterns had the same total activity however the smaller and more contrasting stimulus had elicited much more activity at all layers.

Even the contrasting stimulus (the word "ALO" from group L) being spread into the input space, the attentive behavior is the same one when a contrasting stimulus is focal (inputs G, I, J). Attention is very important, and since the brain is not so much different to any other physical computational system, processing devices are limited. So, the strategies to process information are selecting a part of scene from its complex context and preferentially process it [30, 31, and 32]. How this selection occurs is an unsolved question. Probably there are two kinds of attention: a spaced-based attention [33, 34, 35, 36] that drives a spotlight in the sensorial field to magnify a entity inside it, and a object based attention where the selection mechanism look for the entity per se and not for its placement [36, 37, 38, 39]. An attended visual object is correlated to a gain of activity on neural assemblies where it is projected [40, 41, and 42]. Raizada and Grossberg regard voluntary attention as an important facilitator to low-contrast stimulus because attention would have a similar effect that the contrast exerts by itself upon pop-out stimulus [43]. These authors suggest a model to effort the hypothesis that endogenous and exogenous attention phenomena are inherent to the same functional pathways devoted to perception. The voluntary attention focus could be the related to a top-down enhancement of neuronal activity to desired objects or places and extinction of activity related to distractors upon early and higher stages of visual processing [46, 47, 48, 49]. In a higher level, there are evidences that the activity levels by V4 neurons are set by top-down attention and by stimulus contrast [44]. Several kinds of contextual contrast (luminance, color, bar orientation, movement, etc.) are able to pop-out the stimulus [45].

Koch and colleagues have proposed a theoretical model to attention based on saliency maps [50, 51]. Saliency maps are related to each kind of feature and they enclose the feature contextual contrast related to stimulus. This contrast is the saliency.

Our network could be a model for studying attention system for many reasons. First, our system selects salient stimulus from its context and this saliency could be a bottom-up driving force by natural stimulus contrast or a top-down gain force by voluntary attention mechanisms. The saliency displaces a resulting vector of activity towards itself: energy flows throughout saliency related connections and inhibitory collaterals weaken other possible activities. Second, this property emerges from network structure when it is been processed by different and simultaneous stimuli and the resultant distributed representation is congruent to salient stimulus. The computational capabilities of this model bind the discriminative power to a refined attention mechanism. Third, like discrimination, this behavior is innate, independent of any learning.

References

1. Robert, L.; Goldstone, R. L.: Perceptual learning. Annu. Rev. Psychol. 49 (1998) 585 - 612
2. Dosher, B.A.; Lu, Z.L.:   Perceptual learning in clear displays optimizes perceptual expertise: learning the limiting process. Proc. Natl. Acad. Sci. U S A. 102 (2005) 5286-5290.
3. Dosher, B. A.; Lu, Z. L.: Perceptual learning reflects external noise filtering and internal noise reduction through channel reweighting. Proc. Natl. Acad. Sci. USA 95 (1998) 13988–13993
4. Schiltz, C.; Bodart, J. M.; Dubois, S.; Dejardin, S.; Michel, C.; Roucoux, A.; Crommelinck, M.; Orban, G. A.: Neuronal Mechanisms of Perceptual Learning: Changes in Human Brain Activity with Training in Orientation Discrimination. NeuroImage. 9 (1999) 46–62.
5. Nirenberg, S.; Latham, P. E.: Decoding neuronal spike trains: How important are correlations? Proc. Natl. Acad. Sci. USA. 100 (2003)7348–7353.
6. Thorpe, S.: Localized versus distributed representations. In: Arbib, M. (ed.): Handbook of brain theory and Neural Networks. MIT press, Cambridge. (1995) 549-552
7. Földiák, P.; Young, M. P.: Sparse Coding in the Primate Cortex. In: Arbib M. (ed.): Handbook of brain theory and Neural Networks. MIT press, Cambridge, (1995) 895-898.
8. Tootell, R.B.H.; Dale, A. M.; Sereno, M. I.; Malach, R.: New images from human visual cortex. Trends Neurosci. 19 (1996) 481-488.
9. Engel, A. K.; Roelfsema, P. R.; Fries, P.; Brecht, M.; Singer, W.: Role of the temporal domain for response selection and perceptual binding. Cerebr. Cortex. 7 (1997) 571-582.
10. Katz, D. B.; Nicolelis, M. A. L.; Simon, S. A.: Gustatory processing is dynamic and distributed Curr. Opinion Neurobiol. 12 (2002) 448–454
11. Katz, D. B.; Simon, S. A.; Nicolelis, M. A. L.: Dynamic and Multimodal Responses of Gustatory Cortical Neurons. J. Neurosci. 21 (2001) 4478–4489.
12. Korsching, S.: Olfactory maps and odor images. Curr. Opinion Neurobiol. 12 (2002) 387–392.
13. Theunissen. F. E.: From synchrony to sparseness. TRENDS Neurosci. 26 (2003) 61-64.
14. Linster, C.; Johnson, B. A.; Yue, E.; Morse, A.; Xu, Z.; Hingco, E. E.; Choi, Y.; Choi, M.; Messiha, A.; Leon, M.: Perceptual Correlates of Neural Representations Evoked by Odorant Enantiomers. J. Neurosci., 21 (2001) 9837–9843.
15. Schaefer, M. L.; Young, D. A.;   Restrepo, D.: Olfactory Fingerprints for Major Histocompatibility Complex- Determined Body Odors. J.Neurosci. 21 (2001) 2481–2487.
16. Olshausen, B. A.; Field. D. A.: Sparse coding of sensory inputs. Curr. Opinion Neurobiol. 2004, 14:481–487
17. Gross, C. G.; Desimone, R.; Gattás, R.: Cortical visual areas of the temporal lobe. In: Woolsey, C.N. (ed.): Cortical Sensory Organization - Multiple Visual Areas. 2 ed. The humana press. (1981) 187 – 216
18. Weliky, M.; Fiser, J.; Hunt, R.H.; Wagner, D.N.: Coding of natural scenes in primary visual cortex. Neuron. 37(2003) 703-18.
19. DeWeese, M. R.; Wehr, M.; Zador, A. M.: Binary Spiking in Auditory Cortex. J. Neurosci. 23 (2003) 7940 –7949.
20. Laurent, G.: Dynamical representation of odors by oscillating and evolving neural assemblies. Trends Neurosci.19 (1996) 489-496.
21. Crick, F.; Koch, K.: Constraints on cortical and thalamic projections: the non-strong loop hypothesis. Nature. 391 (1997) 245-450.

22. Shepherd, G.; Koch, C.: Appendix: dendritic electrotonus and synaptic integration. In.: Shepherd, G. (ed.): Synaptic organization of the brain. 3rd Ed.. The Oxford Univ. Press, Oxford. (1990) 439-475.

23. Koch, C.: Computation and the single neuron. Nature. 385 (1997) 207-210.

24. Hilgetag, C. C.; Kaiser, M.: Clustered Organization of Cortical Connectivity. Neuroinformatics. 2 (2004) 353 360.

25. Totoni, G.; Sporns, O.; Edelman, G. M.: A measure for brain complexity: relating functional segregation and integration in the nervous system. Proc. Natl. Acad. Sci. USA. 91 (1994) 5033-5037.

26. Cherniak, C.: Component Placement Optimization in the Brain. J. Neurosci. 14 (1994) 2418-2427.

27. Buzsáki, G.; Geisler, C.; Henze, D. A.; Wang, X-J.: Interneuron Diversity series: Circuit complexity and axon wiring economy of cortical interneurons. TRENDS Neurosci. 27 (2004) 186 – 193.

28. Goodhill, G. J.; Carreira-Perpiñán, M.: Cortical Columns. In.: Nadel, L. (Ed.): Encyclopedia of Cognitive Science, Macmillan Publishers Ltd. (2002) 1 – 9.

29. Anderson, J.A.; Silverstein, J.W.; Ritz, S.A.; Jones, R.S.: Distinctive features, categorical perception, and probability learning: some applications of a neural model. Psychol. Rev. 84 (1977) 413-451

30. Olshausen, B. A.; Koch, C.: Selective visual attention. In.: Arbib, M. A. (ed.): The Handbook of brain theory and neural networks. MIT press, Cambridge. (1995) 837 – 840.

31. Laurent Itti, L.; Koch, C.: Computational modelling of visual attention. Nature Rev., 2 (2001)194-203.

32. Groh, J. M.; Seidemann, E.; Newsome, W. T.: Neurophysiology: Neural fingerprints of visual attention. Curr. Biol. 6 (1996) 1406–1409.

33. Galin, D. Comments on Epstein's Neurocognitive Interpretation of William James's Model of Consciousness. Consciousness and Cognition. 9 (2000) 576–583.

34. Gobell, J. L; Tseng, C-H; Sperling, G.: The spatial distribution of visual attention. Vision Research 44 (2004) 1273–1296

35. Baars, B.: In the theatre of consciousness: global workspace theory, a rigorous scientific theory of consciousness. J Consciousness Studies. 4 (1997) 292-309.

36. Lauwereyns, J.: Exogenous/Endogenous Control of Space-based/ Object-based Attention: Four Types of Visual Selection? Eur. J. Cogn. Psychol. 10 (1998) 41 – 74.

37. Egly, R.; Driver, J.; Rafal, R. D. : Shifting visual attention between objects and locations: Evidence from normal and parietal lesion subjects. J. Exp. Psychol Gen. 123 (1994) 161-177.

38. Iani, C.; Nicolleti, R.; Rubichi, S.; Umihà, C.: Shifting attention between objects. Cognitve brain Res. 11 (2001) 157-164.

39. Abrams, R. A.; Law, M. B.: Random visual noise impairs object-based attention. Exp. Brain Res. 142 (2002) 349–353.

40. O'Connor, D.H.; Fukui, M.M.; Pinsk, M. A.; Kastner, S.: Attention modulates responses in the human lateral geniculate nucleus. Nature neurosci. 5 (2002) 1203-1209.

41. Yantis, S.; Serences, J. T.: Cortical mechanisms of space-based and object-based attentional control. Curr. Opinion Neurobiol.13 (2003)187–193.

42. Pieter R. Roelfsema, Victor A. F. Lamme & Henk Spekreijse Object-based attention in the primary visual cortex of the macaque monkey. Nature. 395 (1998) 376-381.

43. Raizada, R. D. S.; Grossberg, S.: Context-Sensitive Binding by the Laminar Circuits of V1 and V2: A Unified Model of Perceptual Grouping, Attention, and Orientation Contrast. Visual Cognition. 8 (2001) 431-466.

44. Reynolds, J. H.; Pasternak, T.; Desimone, R.:. Attention Increases Sensitivity of V4 Neurons. Neuron. 26 (2000) 703–714.

45. Wolfe, J.: Visual attention. In.: De Valois, K.K. (ed.): Seeing. 2nd ed. Academic Press, San Diego. (2000) 335-386.

46. Desimone, R. Visual attention mediated by biased competition in extrastriate visual cortex. Phil.Trans. R. Soc. Lond. B (1998) 353, 1245-1255

47. Kastner, S.; De Weerd, P.; Desimone, R.; Ungerleider, L. G.: Mechanisms of Directed Attention in the Human Extrastriate Cortex as Revealed by Functional MRI. Science. 282 (1998) 108-111.

48. Kastner, S.; Pinsk, M. A.; De Weerd, P.; Desimone, R.; Ungerleider, L. D.: Increased Activity in Human Visual Cortex during Directed Attention in the Absence of Visual Stimulation. Neuron. 22 (1999) 751–761

49. Usher, M.; Niebur, E.: Modeling the Temporal Dynamics of IT Neurons in Visual Search: A Mechanism for Top-Down Selective Attention. J. Cognitive Neurosci. 8 (1996) 311 – 327.

50. Rutishauser, U.; Walther, D.; Koch, C; Perona, P.: Is bottom-up attention useful for object recognition? IEEE International Conference on Computer Vision and Pattern Recognition. 2 (2004) 37-44

51. Koch, C.; Ullman, S.: Shifts in selective visual attention: towards the underlying neural circuitry. Hum. Neurobiol. 4 (1985) 219-227

AC³ – Automatic Cartography of Cultural Contents

Jean-Gabriel Ganascia

LIP6 - Université Pierre et Marie Curie (Paris VI), 8, rue du capitaine Scott,
75015, Paris, France
Jean-Gabriel.Ganascia@lip6.fr

Abstract. Experiences with e-books show that the principle obstacle to electronic reading is neither the weight, nor the autonomy or the discomfort of reading on screen, but the absence of reference mark which makes it possible to replace the current window of reading in the whole of the book. We present an automatic cartography of electronic documents which constitutes an attempt to facilitate navigation, reading and memorization of contents. It is to automatically build a singular picture which is designed to be associated to our remembering of each document, i.e. to our mental image. This picture corresponds to the cartography of an island. Its shape is build from the document structure; its coloring reflects the affective content of the text extracted by keywords spotting techniques while icons associated to document topics are added to textual legends as in ancient geographical maps.

1 Introduction

Nowadays, electronic versions of almost all classical books are freely available through internet. There exist many open access electronics libraries (e.g. Guttenberg project, Bibliothèque Nationale de France, etc.) where those electronic texts are stored and obtainable. The Google Print project is intended to increase even more the number of those digitalized books accessible on line. Most of the newspapers propose free electronics versions of their papers. More generally, new devices such MP3 players (e.g. Apple IPOD) make everybody able to store and hear a lot of music which is now easily available on the web. Let us note that, for instance, the Apple IPOD has a 60 Gigabytes memory, which roughly corresponds to 1000 audio CD. Similarly, new on demand radio and TV will store 40 hours of TV (and far more of radio). Therefore, many cultural contents (novels, music, photos, videos, etc.) are now accessible through electronic files.

Nevertheless, more the quantity of accessible and storable contents increases, more difficult it becomes to navigate through all those cultural contents and, consequently, to take advantage of all the possibilities opened by new devices. Past experiences [1, 7] show that the principal impediments are due to the difficulty to materialize virtual contents which appear very abstract. As a consequence, people are lost in electronic libraries, whatever they contain, texts, music, photographs or movies. For instance, the failure of e-books [12, 13] which were build a few years ago to ease the electronic reading is neither due to their cost, nor to their weight, their autonomy or the discomfort of reading on screen, but to the lack of reference mark which makes it possible to replace the current window of reading in the whole of the book. This

P.P. Lévy et al. (Eds.): VIEW 2006, LNCS 4370, pp. 253–263, 2007.

absence makes people lost which requires considerable efforts to navigate in the text and to memorize the content. The e-book designers introduced a lift which does not appear to be sufficient. Even if the position in the page and the thickness of the section in classical books are not always consciously perceived, they provide precious information which appears to be lost in e-books.

Our goal here is to mitigate these deficiencies and to improve existing e-reading interfaces by "embodying" virtual contents into a 2D picture where each picture elements, i.e. pixels, correspond to text fragments viewed at different scale: sentence, paragraph, section, chapter, etc. To achieve this "embodiment", we propose a visualization tool which automatically builds a map helping people to find their way in virtual worlds. More precisely, it is to cartography the structure of electronic documents making navigation through them easier. The key idea is to project the abstract structure of the book into a spatial imaginative territory which can be represented on an automatically generated map. It will then be possible for readers to plan new travels through books or multimedia contents on this map and to keep tracks of their past travels. Since each book fragment – sentence, paragraph, section, chapter ect. – is associated to a pixel, the general map that organizes all those book pixels constitutes a "pixelisation" of the book.

Apart the introduction and the conclusion, the paper is divided into three main parts. The first is dedicated to the depiction of the main idea, which is to automatically draw a map reflecting the document structure. The second presents some technical aspects of the shape generation process while the third will describe augmentations of generated maps through colors, icons and active traces.

2 Memory Islands

The main idea of this paper comes from the old *arts of memory* [4, 11, 19] which were based on the spatial representation of contents. It means that in the Antiquity and in the Middle Age, people developed a way to improve their memory abilities by localizing things in virtual architectures. We transpose that idea to the content of electronic book: our hypothesis is that the main obstacles to electronic reading could be solved if we were able to place what we call the "Reading Space" by analogy to the Bolter's "Writing Space" [2] in virtual territories.

In other words, to "embody" the virtual content of books, we anchor it on an automatically generated territory. In a way, it is to represent on a plan, i.e. on a 2D space, the book content that appears at first sight to be linear. It corresponds to an increase of dimensions, which is quite unusual in information visualization, since generally the goal is to reduce data dimensions [15, 16, 17]. The adopted solution is to map each book onto a small deserted island, because uninhabited islands are known just by sailors who usually make simply the turn of it, and therefore know and name only their coast, i.e. a 1D space, without worrying about their surface.

These imaginary territories are designed to strike imagination and to remain anchored in our memory. Consequently, they must as much as possible be distinguished the ones from the others. Moreover, it would be preferable that they reflect the structure of the books to ease navigation. The map generation program endeavor to do it by printing forms as diversified as possible and while exploiting the colors. More precisely, the program builds a realistic map of a territory on which it is possible to have an overview of contents. This cartography takes advantages of historical studies of old maps [8], for instance, the introduction of legends, texts and colors, and, more amazingly for us, the insertion of icons associated with the content. As an example, the map of North Africa presented in figure 1, shows flags, elephants, lions, camels, castles and caravans, that are drawn to localize places where visitors could find elephants, lions, camels, cities and caravans. Moreover, always in this map, the red sea is colored in red. Even if they are not realistic, the icons and colors help to memorize the map. They strike our imagination in a way which facilitates our internal visualization and consequently our remembering. Similarly, the artificial maps our program build are enriched by colors and icons correlated to the topics under consideration. It is not to summarize the content, but to stimulate our memory with an easy to remember picture.

Fig. 1. An old map of Africa ("Aphrica" on the flag) attaching colors and icons to territories

3 Shape Generation

E-document structure can be assimilated to an ordered weighed tree, i.e. to an ordered tree of which each node is associated to a number corresponding to the weight of the corresponding part, for instance to the number of pages for texts or to the required space for other multimedia contents i.e. for MP3 files. Then, a shape is generated from this tree. Let us recall that the goal is to generate singular forms attached to each particular e-document that have both to be easy to retain in our memory and to reflect the document structure, which facilitates the navigation. Therefore, the goal of the shape generation process is to differentiate forms as much as possible.

More precisely, the shape generation algorithm considers each weighted node of the e-document structure, i.e. each part of the document, as a sector of a disk, i.e. as an angle associated with a radius. The angle corresponds to the proportion of the section in the whole book while the radius is computed taking into account the size of the section and its level. Moreover, some blank pages are virtually considered to separate sections, which cut out the coast of the artificially generated island. For more details, about the algorithm, see [6].

As an illustration, let us consider a given book [5] of which structure is given in its table of contents (see figure 2).

Table des matières

Microbe ou esprit?	7
«La peur»	7
Esprits, êtes-vous encore là?	10
Echec du positivisme ordinaire	17
Retour au fétichisme	23
Des sciences de l'esprit	27
Déroute de l'intelligence	31
Les deux infinis...	33
Le mur de la complexité	36
L'âge labyrinthique	46
Lumières	55
Connais-toi toi-même	73
Imagerie cérébrale	84
De l'imagerie moderne à la lunette de «maître puce»	90
Mathématisation du psychisme	94
Les Golems modernes	111
Linceuls de nos mémoires	119
Information	120
Nombres informatiques	123

Fig. 2. Table of content showing chapters and sections associated with page numbers

From this table of content, it is possible to generate a weighted ordered tree (see below, figure 3) where chapter and section weights correspond to number of pages they contain.

Fig. 3. Ordered tree corresponding to the book structure extracted from the table of content given in fig. 2

Then, the shape generation algorithm builds a map of an imagined island corresponding to this structure (see figure 4).

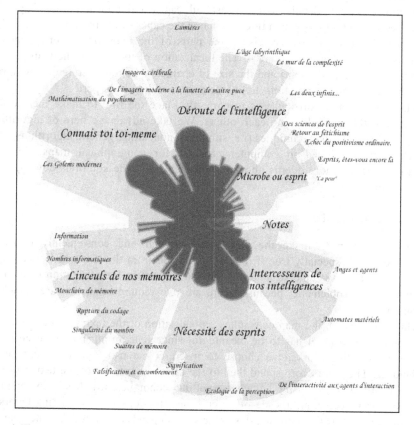

Fig. 4. The map generated from the table of content given in figure 3 using two levels of color

The important point is to make the reader able to localize all part of the book or of the library on the map. It is also required to have a specific form which is easier to remember and where the reader can attach souvenirs to each subpart, as we do on a geographical map.

4 Augmentations

To make the map more singular and, consequently, to ease the memorization, the program insert texts, colors and icons. Texts correspond to titles of chapters, sections and subsections. Colors are associated to affective content of different book subparts while icons match topic categories.

4.1 Texts

As in traditional maps, texts are associated to places. Since places represent parts of books, texts name parts of books, i.e. chapters, sections, subsections etc. More precisely, the coast corresponds to the fine structure of the book, i.e. to the subsections. So, all details of the coast, all capes and almost all the bays, are generated by subsections which title may name the capes of the bays on the map. The section and chapter titles may also name parts of the map, but it corresponds to internal parts, as shown in figure 4. The localization of the text on the map (i.e. the eccentricity of its placement) and the size of characters help to distinguish the levels (i.e. chapter, section, subsection etc.) of the corresponding book part.

The only practical difficulty is to correctly place the title to make them easily readable, without overlapping each others. We take advantage here of algorithms based on optimization techniques (i.e. simulated annealing, genetic algorithms, etc.) which were developed in computer graphics for geographical maps. See, for instance [12] for more details about the problems and the solutions developed to solve them.

4.2 Coloring the Map

Colors may help to make the map more singular and consequently easier to remember. It is especially true if they are extracted from the text which means that they reflect, in a way, the structure of the text. However, the attribution of color to pixel may be problematic, because, there are two requirements that are somehow antagonistic.

On the one hand, the map constitutes a pixelisation of the text; as a consequence, the color of each pixel has to be meaningful since it is an attribute of each pixel. Following existing works, we choose to insert colors using affective computing techniques. The colors correspond then to the affective tonality of the text, which is automatically extracted using keywords spotting techniques. It is based on cognitive science works describing a scale of seven basic emotions and associating colors to each of one [3, 9, 18]. Obviously, this association depends on culture, but the range of color could easily be changed and a legend associated to the map explains the meaning of each color.

On the other hand, to ease map reading, it is necessary that the colors obey a certain number of classical coloring rules stated for instance by Edward Tufte [15, 16, 17]. For instance, the contrast of color has to be low; it is not possible to have bright color closed to each other, etc. We have to introduce those coloring rules and to specify, in each case, a legend where the meaning of each color is designed.

Affective tonality is computed at different levels of granularity – sentence, paragraph, sub-section, section, chapter etc. [9]. Colors are then affected to each of the map sector, depending on the considered level of granularity. In other words, the resulting map is obtained by the superposition of overlapping drawings corresponding each to a level of granularity. For instance figure 4 shows the superposition of two levels of granularity of which, for the sake of clarity, color is manually attributed. More precisely, the kernel of the island reproduces the form of the island with the same shape generation algorithm as mentioned before, but with different parameters, making the form a little bit different.

4.3 Adding Meaningful Icons

Moreover, to guide the readers both during his travel of after, memorizing what he had seen, small drawings are placed onto the map. This insertion is similar to an ancient practice of cartographers. Usually, icons refer to general idea by analogy or metonymy. Once a table of conventional icons is established, a content analysis helps to categorize the meaning each part of the text and to associate it a relevant icon. Moreover, it could also be possible for the reader to manually add its own icons corresponding to bookmarks, to notes or to reminders. For instance, the figure 5 shows icons associated to a map automatically generated from the structure of a talk about memorization strategies using artificial maps.

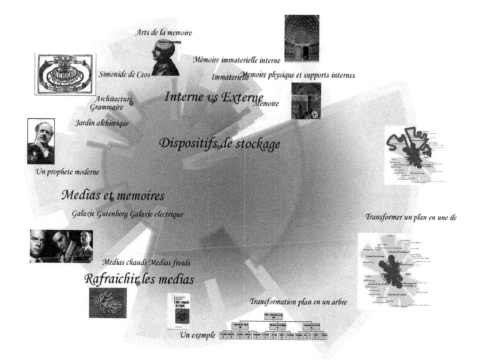

Fig. 5. Artificial map enriched with icons

All the icons are manually introduced on the map; they facilitate memorization of the content. The future step is to generate automatically those icons using a search engine, for instance "google image".

Let us note that the map presented figure 5 has been drawn to accompany a power point presentation. In the future, it would be a challenging application for artificial map, to support presentation, helping the participants both to follow the structure of the talk and to memorize its content.

4.4 Transforming Map and Tracing Reader Travels

To finish, it is possible transform the map according to the needs of the reader which can focus on such or such part of the map with the help of a zooming mechanism. It is also possible to access any particular page, or section, by clicking on the map, since each point of the map is linked to the original text or to MP3 files.

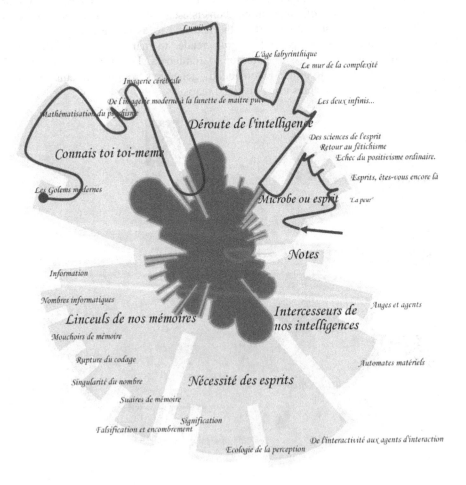

Fig. 6. Tracking the reader course on the map

Last of all, users' courses may be tracked with a colored ribbon drawn onto the map, which remind the history of the reading. For instance, the reading course of a reader starting the book presented in figure 2 at the beginning and reading linearly until section entitled "*Les Golems modernes*", is presented figure 6. It is also possible to compute the width of the ribbon with respect to the time spent to read the different page. It can be very useful, since it may help to distinguish jumps, simple browsing and in-depth reading.

5 Evaluation

The last point concerns the evaluation of the efficiency of the Memory Islands. Let us first note that most studies concerning e-books have focused on the speed and comfort of reading. However, experiments have shown that the problem does not lie in the ergonomics of reading, which appears to be satisfactory, but on the effectiveness of the reading process. In other terms, reading on an electronic support, i.e. on a screen, is relatively fast and comfortable: it does not provoke tiredness, although the vertical position may cause neck stiffness. New platforms such as TablePCs or Pocket PCs will prevent it. The three main issues with the digital reading terminal are:

1. difficulties with spatial and temporal navigation;
2. lack of landmarks for memorizing and understanding the structure of the document;
3. the underlying model of the book: Codex or Scroll, both having important advantages and pitfalls.

The visualization tool we design and experiment within the framework of the Memory Islands paradigm should help memorization, at least if our hypothesis holds that the main obstacles to reading on electronic support are due to the three cited issues. More precisely, the goal of memory islands is to implement visual representations, to add landmarks helping memorization and to implement a new model of book, which will improve on both Codex and Scroll.

Our goal, through evaluation, is provide some empirical evidence confirming our hypothesis. The evaluation makes use of TablePCs and couples them with existing reading software. It is to validate the reading efficiency improvement, due to memory islands through controlled and longitudinal evaluations. This validation will relate to various reading situations (technical reading, reading of newspapers, continuous reading), on various populations (young students, students, and elder people) and it will be carried out in of two phases: immediately after the readings (controlled evaluation) and on a longer term – one week or one month after the readings. Since the goal is not to study reading speed, but efficiency of reading through memorization, we have to test the ability to remember what people have read with and without the Memory Islands visualization software.

More precisely, the experimental protocol will take into consideration two populations: young students and elder people. Each of those populations will be divided into two subsets; the first will have to read e-books with the Memory Islands visualization tool, the second without. For each of those populations, we define two reading situations: technical contents and continuous reading for students, newspapers and continuous reading for elder people. Then, each of the reader will have to answer two questionnaires about what he had read, the first will be given just after reading, i.e. one day after, the second ten or twenty days after. Since our hypothesis is that the use of Memory Island visualization tools will improve memorization, the result of this experimentation should confirm or disconfirm our hypothesis.

As a consequence, the result of this study should substantially improve the design of effective reading tools for screens. Furthermore, it will show specific needs and capabilities of the different populations and provide specific improvements for each of these segments.

6 Conclusion and Perspective

A first prototype of the map generation system has been achieved for texts. It is coupled with Mozilla navigator which contains a native SVG viewer. The input texts have to be written in HTML. Then the structure of document is extracted and the corresponding shape generated. The map is drawn in a small floating window which may be enlarged to the entire screen. Up to now, colors and icons are manually inserted. However, we plan a new version with automatic coloring based on affective computing techniques.

Moreover, our prototype is designed to be coupled with an e-book reader. Nevertheless, we plan a new version dedicated to manage TV content on intelligent TV decoders.

The system is currently under experimentation. The protocol has to test the efficiency of reading with such an "embodiment" of content. Let us note that, contrary to most of e-reading experiments, our goal is not to evaluate the speed of reading, but its efficiency in terms of memorization and knowledge acquisition. More precisely, it has to give students e-books to read and to evaluate their ability to memorize contents with and without the content map, just after reading and a week later.

Lastly, let us note that even if the first and main goal of the project is to improve efficiency of reading and to facilitate navigation through virtual contents by making them more concrete, one of its side effects is to make able to visualize users' travels through documents. Therefore, it could be used to facilitate user centered design of electronic books or multimedia by visualizing users' feedback.

References

1. Back Maribeth, Cohen Jonathan, Gold Rich, Harrison Steve, "Speeder Reader: An experiment in the Future of Reading", *Computers and Graphics*, Vol. 26 (3), June 2002
2. Bolter J. D., *The writing space : the computer, hypertext and the history of writing*, Paperback 1991
3. Boucouvalas A, "Real Time Text-to-Emotion Engine for Expressive Internet Communications", *Being There: Concepts, effects and measurement of user presence in synthetic environments*, G. Riva, F. Davide, W.A IJsselsteijn (Eds.), Ios Press, 2003, Amsterdam, The Netherlands
4. Carruthers Mary J., *The Book of memory : a study of memory in medieval culture* Paperback 1992

5. Ganascia J.-G., *2001, l'odyssée de l'esprit* – Flammarion (Collection essais) 1999.
6. Ganascia J-G, "RECIT : représentation cartographique et insulaire de texts", *proceedings of CIFT 2004* (Colloque International sur la Fouille de Texte), June 2004, la Rochelle, France
7. Harrison, S. R.; Minneman, S. L.; Back, M. J.; Balsamo, A. M.; Chow, M. D.; Gold, R.; Gorbet, M.; MacDonald, D. W. "The what of XFR: eXperiments in the future of reading". *ACM Interactions.* 2001 May/June; 21-30.
8. Jacob C., *L'empire des cartes*, Albin Michel 1992
9. Liu H., Selker T., Lieberman H. (2003). "Visualizing the Affective Structure of a Text Document. *Proceedings of the Conference on Human Factors in Computing Systems", CHI*
10. Prendinger H, Ishizuka M., *Life-Like Characters Tools, Affective Functions, and Applications*, series Cognitive Technologies, Springer 2004
11. Rossi Paolo, *Clavis universalis: Arti della memoria e logica combinatoria da Lullo a Leibniz*, Il Mulino 1983
12. Ryall Kathy, Marks Joe, Shieber Stuart, "An interactive constraint-based system for drawing graphs", *Symposium on User Interface Software and Technology*, Banff, Alberta, Canada, Proceedings of the 10th annual ACM symposium on User interface software and technology, 1997, pp. 97 - 104
13. Shilit B., Morgan P., Golovchinsky G., "Digital Library Information Appliances", Digital Libraries 1998, Pittsburgh USA, ACM
14. Shilit B., Morgan P., Golovchinsky G., Tanaka K., Marschall C., "As We May Read, The Reading Appliance Revolution", Computer, IEEE, January 1999
15. Tufte Edward, *The Visual Display of Quantitative Information*, Graphics Press, 2nd edition, 2001
16. Tufte Edward, Visual *Explanations: Images and Quantities, Evidence and Narrative*, Graphics Press, February 1997
17. Tufte Edward, *Envisioning Information*, Graphics Press, 1990
18. Valdez, P., Mehrabian, A. *Effects of color on emotions.* Journal of Experimental Psychology: General, 123, 394-409. (1994).
19. Yates F., *The Art of Memory*, Penguin Books, 1969

Evaluation of the Mavigator

Mariusz Trzaska[1] and Kazimierz Subieta[1,2]

[1] Polish-Japanese Institute of IT, Koszykowa 86, Warsaw, Poland
[2] Institute of Computer Science PAS, Ordona 21, Warsaw, Poland

Abstract. Mavigator is a graphical querying and browsing tool dedicated to naïve users (computer non-professionals). It allows them to retrieve and analyse information from various data sources, in particular, from object-oriented and XML-oriented databases, providing a corresponding wrapper is implemented. Mavigator key concepts related to information retrieval include: intensional navigation, extensional navigation, and persistent baskets for recording temporary and final results. A basic problem related to end-user visual interfaces concerns final presentation and further processing of retrieval results. The Mavigator Active Extensions (AE) module allows the programmer to extend ad hoc the existing core functionalities through a program written in C#. Thus the retrieved data can be presented and analysed in any conceivable visual form. Another novel feature of Mavigator is Virtual Schemas, which make it possible to customize database schema, in particular, changing some names, adding virtual associations or hiding some classes. Virtual Schemas allow creating a customized version of an existing database schema and navigate within the database according to this schema. In this paper we present results of usability tests, which have been conducted within three independent student groups. The results of the tests suggest improvements to the prototype.

1 Introduction

According to [1] and [2], the application's evaluation process has four main aims:

- Discovering major problems that could result in a human error or lead to frustration of the user,
- Reducing training time,
- Increasing performance and efficiency,
- Improving user's satisfaction from using the software.

The usability of an application can be defined as an ability to satisfy user's needs related to the application. As noted in [3] there are two fundamental approaches to usability:

- By principles. This means choosing such usability principles, which will be adequate for a particular kind of an application and user,
- By evaluation. This approach requires evaluation by the users, which means that the whole application (or at least some part of them) must be developed. The easy part of the method is criticizing current solutions. The hard part is deciding what to change to improve it.

P.P. Lévy et al. (Eds.): VIEW 2006, LNCS 4370, pp. 264–277, 2007.

We believe that developers of applications have to combine two above approaches. During the entire application's creation process, all known usability principles must be taken into consideration. Still, after developing a beta version, the evaluation with users must be eventually conducted. In this paper we present the research results along this line.

In [4] Plaisant distinguishes four thematic areas of evaluation:

- Controlled experiments comparing design elements. Such studies should compare specific widgets and mappings of information to a graphical display,
- Usability evaluation of a tool. Such studies should provide feedback on the problems that the users encounter with a tool and should explain how the designers have to refine the design,
- Controlled experiments comparing two or more tools. These studies usually try to compare a novel technique with the state of the art.
- Case studies of tools in realistic settings. This is the least common type of studies. The advantage of case studies is that they report on users in their natural environment doing real tasks, demonstrating feasibility and in-context usefulness. The disadvantage is that they are time consuming to conduct, and the results may not be replicable.

Our first prototype SKGN has been used in the European project ICONS. It was informally evaluated during using for the Structural Fund Projects Portal implemented at the top of ICONS. Hence we have some informal response from the users, generally very positive. However, we need more formal evaluation, thus we have decided to conduct usability evaluation of the tool (the second area of evaluation). The procedure is generally enough to utilize them in evaluating almost every kind of graphical user interfaces, including systems employing pixellisation paradigm.

To fully understand the evaluation process some general information regarding the Mavigator must be presented. Hence, after presenting related work (Section 2), next sections briefly discuss key concepts of the Mavigator: information retrieval capabilities (Section3), Active Extensions (Section 4) and Virtual Schemas (Section 5). Section 6 is dedicated to the evaluation itself. Section 7 concludes.

2   Related Work

Related solutions can be analysed from three points of views: methods of modifying application's functionalities, the way of information retrieval and utilization of database views. However due to the limited space we will focus on the second topic only (more information can be found in [5, 6]).

Roughly, visual metaphors to information retrieval can be subdivided into two groups: based on graphical query languages and graphical browsing interfaces. The subdivision is not fully precise because many systems have features from both groups. An example is Pesto [7] having possibilities to browse through objects from a database. Otherwise to Mavigator, the browsing is performed from one object to a next one. For instance, the user can display a Student class object, but to see another student, he/she needs to click next (or previous) button and replace current visualization. Besides browsing, Pesto supports quite powerful query capabilities. It utilizes the query-in-place feature, which enables the user to access nested objects, e.g. courses of particular students, but still in the one-by-one mode. Another advantage concerns complex queries with the use of existential and universal quantification. Such complex features may however compromise usability for less professional users and some kinds of retrieval tasks.

Typical visual querying systems are Kaleidoscape [8] and VOODOO [9]. Both are declared to be visual counterparts of ODMG OQL thus graphical queries are first translated to their textual counterpart and then processed by an already implemented query engine. The first one uses an interesting approach to deal with AND/OR predicates. We find it very useful and intuitive thus we have adopted it to our metaphor.

A typical example of a browsing system is GOOVI [10]. Unfortunately, selecting of objects is done via a textual query editor. A strong point of the system is the ability to work with heterogeneous data sources.

Another interesting browser is Watson [11], which is dedicated to Criminal Intelligence Analysis. It is based on an object graph and provides facilities to make various analyses. Some of them are: retrieving all objects connected directly/indirectly to specified objects (i.e. e. all people, who are connected to a suspected man), finding similar objects, etc. Querying capabilities include filtering based on attributes and filter patterns. The latter allow filtering links in a valid path by their name, associated type, direction or a combination of these methods. The manner of work is similar to the metaphor that we have called extensional navigation.

Browsing systems relay on manual navigation from one object to a next one. During browsing the user can read the content of selected objects. Browsing should be an obligatory option in situations when the user cannot define formally and precisely the criteria concerning the search goal.

3 Information Retrieval Capabilities

Mavigator is made up of three metaphors utilized for information retrieval: intensional navigation, extensional navigation and persistent baskets. The user can combine these metaphors in an arbitrary way to accomplish a specific task.

Intensional and extensional navigation are based on navigation in a graph according to semantic associations among objects. Because a schema graph (usually dozens of nodes) is much smaller than a corresponding object graph (possibly millions of nodes), we anticipate that intensional navigation will be used as a basic retrieval method, while extensional navigation will be auxiliary and used primarily to refine the results. Next subsections contain short description of the methods (more information can be found in [12]).

3.1 Intensional Navigation

Intensional navigation utilizes a database schema graph. Figure 1 shows a window containing a database scheme graph of the Northwind sample. The graph consists of the following primitives:

- Vertices, which represent classes or collections of objects. With each of them we associate two numbers: the number of objects that are marked by the user and the number of all objects in the class,
- Edges, which represent semantic associations among objects (in UML terms),
- Labels with names of association roles. They are understood as pointers from objects to objects (like in the ODMG standard, C++ binding).

The user can navigate through vertices via edges. Objects, which are relevant for the user (candidates to be within the search result) can be marked, i.e. added to the group of marked objects. There are a number of actions, which cause objects to be marked:

- Filtering through a predicate based on objects' attributes.
- Manual selection. Using special labels it is possible to mark particular objects manually. It is especially useful when the number of objects is not too large and there are no common properties among them.

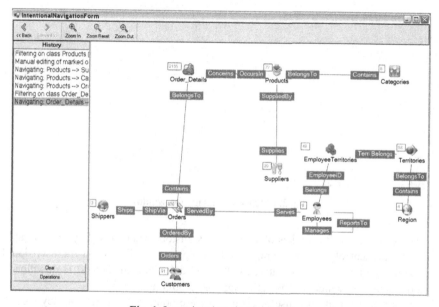

Fig. 1. Intensional navigation window

- Navigation from marked objects of one class, through a selected association role, to objects of another class. An object from a target class becomes marked if there is an association link to the object from a marked object in the source class.
- Basket activities (see further).
- Active extensions (see further).

Intensional navigation and its features allow the user to receive (in many steps but in a simple way) the same effects as through complex, nested queries. Integrating these methods with an extensional navigation, manual selection and other options supports the user even with the power not available in typical query languages.

3.2 Extensional Navigation

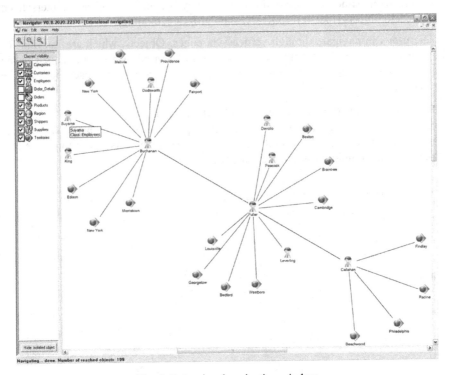

Fig. 2. Extensional navigation window

Extensional navigation takes place inside extensions of classes (Figure 2). Graph's vertices represent objects, and graph's edges represent links. When the user double clicks on a vertex, an appropriate neighbourhood (objects and links) is downloaded from the database, which means "growing" of the graph.

Extensional navigation is useful when there are no common rules (or they are hard to define) among required objects. In such a situation the user can start navigation from any related object, and then follow the links. It is possible to use basket for storing temporary objects or to use them as starting points for the navigation.

Baskets are persistent storages of search results. They store two kinds of entities: unique object identifiers (OIDs) (seen as object labels) and sub-baskets. The hierarchy of baskets is especially useful for information categorization and keeping order. During both kinds of navigation it is possible to drag an object (or a set of marked objects) and to drop them onto a basket. The main basket is assigned to a particular user. At the end of a user session all the baskets are stored in the database.

3.3 Baskets

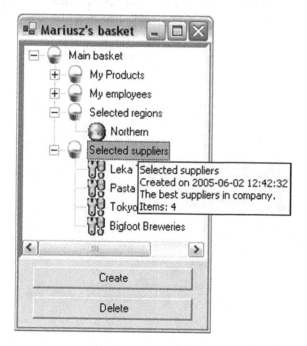

Fig. 3. Basket visualization

Baskets allow storing selected objects in a very intuitive and structured way. Navigation can be stopped at any time and temporary results can be named, stored and accessed at any time.

4 Active Extensions

Mavigator already employees some information retrieval metaphors (see section 3), which are powerful and yet easy-to-use, so we have decided to provide a way to add new functionalities operating only on a query result. The approach does not complicate the entire application, but guarantees sufficient flexibility. Thus our solution requires collaboration of an end user with a programmer who will write the code accomplishing the required functionalities.

The current Mavigator prototype uses Microsoft C# as a language for Active Extensions. A programmer is aware of the Mavigator metadata, which allows him/her to write a source code of the required functionality in C#. Writing the Active Extension source code is done in the Mavigator's special editor. Once a programmer compiles the code, a particular Active Extension is ready to use (without stopping Mavigator). Then the end user is supported with one click button causing execution of the written code. It processes the query result (see Section 3.1) or objects recorded in a user basket (see Section 3.3). The functionality of such programs is unlimited and can be a simple one (i.e. calculating average value of selected attributes), through objects exporters (i.e. to XML format), up to more complicated ones like Active Projections.

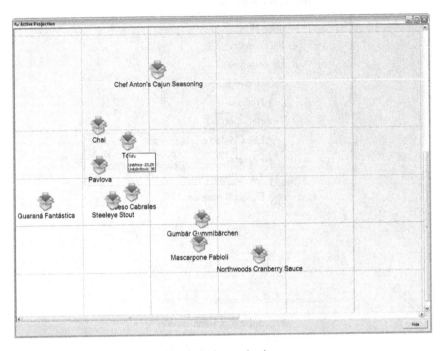

Fig. 4. Active projections

Active Projections (Fig. 4) allow visualizing a set of objects where the position (x and y coordinates) of each of them is based on value of particular objects' attributes. Current implementation uses two axes (2D), which allow visualizing dependencies of two attributes. Besides the visual analysis of objects dependencies it is also possible to utilize projections in more active fashion. Object taken from a basket can be dropped on a projection's surface, which causes right (based on attributes values) placement. It is also possible to perform a reverse action: drag an object from the surface onto a basket, which causes recording the object in the basket. More information can be found in [12].

5 Virtual Schemas

A virtual schema (a view) is a mapping of a database schema according to needs of the current user. A virtual schema exists as a definition only, no physical mapping of data is performed. According to the fundamental transparency requirement, the user uses a virtual schema in the same manner as an original database schema. Virtual schemata have low resource demands, thus Mavigator is capable to support user's work with many virtual schemas in the same retrieval session.

Generally, Virtual Schemas allow creating customized database views consisting of the following elements:

- Virtual associations, which reflect any dependencies among classes,
- Virtual attributes, which describe objects' properties.
- Classes, which are counterparts of physical collections from the database.

More detailed discussion of the topic can be found in [6].

6 Evaluation

Mavigator's evaluation has been conducted by the 6 subjects in three groups (because of problems with scheduling). All of them were students from the Polish-Japanese Institute of Information Technology (different departments). All surveys were in Polish and based on [13, 14].

6.1 Procedure

At the beginning, subjects have to answer questions about their experience and knowledge. All answers were from range 0 (lack of knowledge) to 10 (expert). Below we have enumerated all of them (question 1.1 is just an id of the subject):

1.2   Object-oriented database concepts (classes, associations, etc.),
1.3   Microsoft Access experience,
1.4   Textual query language experiences (i.e. SQL, OQL),
1.5   Visual information retrieval system experience,
1.6   Programming language experience.

Then, all subjects were trained on the Mavigator prototype. The training program consisted of short introduction with description of motivations behind Mavigator, discussion of Mavigator's key concepts and detailed instructions on using the prototype. This includes demonstration of using particular techniques like filtering, navigating, etc.

After the training, subjects have to find answers for following query questions (working with the "Northwind" database mentioned earlier):

1. Find all employees with the first name "Robert".
2. Find all employees with the first name "Robert" or "Nancy".
3. Find the name of an employee who manages the territory "Cambridge".
4. Find names and prices for all products from supplier "Leka Trading" where prices are more than $18.00.
5. Find employees who served orders sent to Mexico and manage territories located in the "Northern" region.
6. Find average price of the products sent to the Mexico.
7. Find the cheapest product among those ones, which are in stock.
8. Analyze the "Seafood" products, to find out which of them are not so much on the stock.

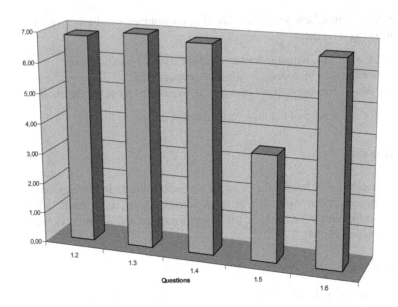

Fig. 5. Average information about subjects

For each query question, each subject has to fulfil survey with the following items (2.1 is an id of the subject, 2.2 is the number of the query question):

2.3   Difficulty of the query question – from 1 (easy) to 10 (very hard),
2.4   Success - in percents,
2.5   Time to completion – in minutes,
2.6   Remarks.

At the end of evaluating, after solving all query problems, each user answers summary questions about the metaphors, prototype, etc:

3.1   Comprehensibility – from 1 (confusing) to 10 (clear),
3.2   Ease of use – from 1 (difficult) to 10 (easy),
3.3   Speed of use – from 1 (slow) to 10 (fast),
3.4   Performance – from 1 (slow) to 10 (fast),
3.5   Overall satisfaction – from 1 (terrible) to 10 (wonderful),
3.6   Remarks,
3.7   Ideas for improvement,
3.8   Ideas for new Active Extensions.

6.2   Results

Figure 5 shows average values given by subjects describing themselves. Bars refer to the questions enumerated in section 6.1 According to the answers, all subjects have been quite familiar with database concepts (question 1.2), MS Access (question 1.3), textual query languages (question 1.4) and programming language (question 1.6).

However, most of them were not familiar with graphical user interfaces (question 1.5). From the testing point of view, maybe it would be better to find subjects less familiar with these topics. However, we have to take into account that some of the answers could be a little bit exaggerated. This might be caused by the fact that subjects were students (and the administrator of the experiment was their teacher), who should be familiar with the mentioned terms.

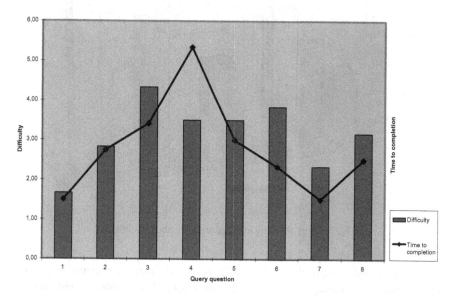

Fig. 6. Chart showing dependence between questions' difficulty and time for completion

Figure 6 presents chart showing two kinds of information (for each query question):

- Average difficulty of the query questions (bars),
- Average time to completion (line).

As it can be seen, queries, which has been judged as harder, take more time to complete (except the fourth one). Also, almost all queries (without one – for one subject), have been completed in 100 percent (not shown on the chart). The shortest time to complete was 1.5 min. (for number 1 and 7), the longest one was 5.5 min. (for number 4) and an average time to complete was about 2 minutes and 45 seconds. According to the subjects' answers, the harder question was number 3 and then number 6. However, even the hardest ones, still have been judged as easier than medium (less then 5 in the 10 degrees scale).

Figure 7 shows average answers (to the questions from section 6.1), given by the subjects, about Mavigator's prototype. Roughly speaking, all answers are positive (more then 5 in the tenth degrees scale).

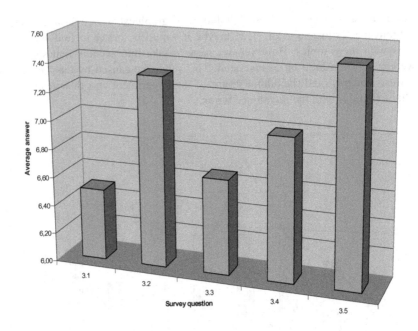

Fig. 7. Chart showing average answers about Mavigator

Comprehensibility (question 3.1), which is a very important factor in graphical user interfaces, has average value 6.5 with minimum at 4 (one subject) and maximum at 8 (also one subject). Most subjects give 7/10.

Another important factor, which has been judged by the subjects, is ease of use (question 3.2). In this case, marks vary from 6, through 7 and 8 up to the 9. Average value was 7.3/10, which is quite high.

Speed of use (question 3.3), which is different from speed of working (perform-ance – 3.4), describes things like the way of using options, running actions, etc. This factor is related in some way to the comprehensibility, and has achieved average value about 6.6/10.

Speed of working (question 3.4) is not so important in case of prototypes. More-over performance of the information retrieval tools (including Mavigator) is tightly connected with the performance of the data source. The current prototype works with the ODRA experimental database server, which has not been optimized yet. However, in this field Mavigator has also gotten quite a good result: 7/10.

The last factor, which has been assessed, overall satisfaction of the user. This is very subjective judgment and has big impact on the user's decision: do I really want to use the tool? Fortunately, Mavigator's users give them average mark about 7.5/10 with minimum at 7 and maximum at 9. Such a result bears good testimony to the ideas and metaphors behind Mavigator:

6.3 Users' Remarks

Subjects have the right (and they have been encouraged to) to make comments and remarks. All of them have been carefully analyzed. Most of them were related to the graphical user interface (and have been taken into account) and not to the metaphors itself:

- Items' Lists. One subject noticed that a list of attributes during defining a single predicate is not sorted any way. As a result all lists (of items of any kind) appearing in the prototype have been analyzed and sorted alphabetically.
- Showing Objects. Another subject gave attention that one of the most frequently performed operations is showing marked objects. Thus, it would be nice to have an opportunity to perform it very quick. Hence, a new way of showing marked objects has been added: when a user double clicks on the class icon, a list of marked object is shown.
- Showing Basket. Next problem was connected with showing basket's window. When the window has been shown, and then another window covered them, choosing Show basket from the menu, has no effect. It has been fixed by bringing the basket's window to the front (after selecting appropriate option form the menu).
- Working with a Single Predicate. When a user defines a single predicate (for filtering some objects), there is a necessity to enter some values. One of the subjects suggests that application should prompt values of the selected attributes read from existing objects. The idea is obviously good. However, its implementation could lead to serious performance problems. This is caused by the fact that prompting values requires reading all objects from the database (or at least all values of the particular attribute). This is technically possible, but the performance overhead makes it questionable.
- Marking objects with filtering. During filtering objects, one of the subjects reported that filtering system does not work. After short investigation it comes out that the user is filtering from 0 marked objects, which leads to marking 0 objects. As a result, a dedicated message has been added, which informs the user about the situation.

6.4 Procedure Remarks

As mentioned previously, Mavigator's evaluation procedure is quite general and could be utilized in many different kinds of graphical information retrieval/analysis tools. In particular, the procedure, without a big effort, could be tailored for evaluating systems which employs pixellisation paradigm. Following items contain a brief discussion of such utilization:

- The first point of the procedure (questions from 1.2 – 1.6) collects information about the subject. Thus, such information are useful during evaluation all kinds of systems. Based on the answers we can judge the results of the entire evaluation.
- Next task included training of the prototype which is necessary in case of all new systems. After that subjects have to find answers to some problems. In case of Mavigator it was some kind of data related queries. In case of other analysis tools it would be other data related activities i.e. finding some relations among information, etc.

- Questions from point 2 give information about difficulty of the problem, time to completion, etc (they should be answered for each query problem). All of them could be applied with minor changes.
- And the last group of questions (3.1 – 3.8) stores some overall opinion about the system. Such questions are necessary during evaluating all kind of computer systems.

7 Conclusions and Future Work

We have presented the Mavigator and its evaluation. Conducted surveys satisfied two important evaluation's targets:

- Allowed improving the prototype,
- Proved rightness of the Mavigator's ideas and metaphors.

Overall high marks, issued by the prototype's users confirm Mavigator's high usability and easy-in-use.

The prototype offers new quality in two main areas. The first one is extending existing application's functionalities. Active Extensions, which use fully-fledged programming language, make it possible to create any kind of additions to Mavigator's core functions. The second area contains Virtual Schemas, which allow creating customized version of existing database schema.

As a continuation of our research, we have started a work related with connecting the Mavigator with the eGov-Bus virtual repository. This data source is developed as a part of the EC project called eGov-Bus[1].

We also plan investigation on adding new functionalities, which will make our system more powerful and easy-to-use.

References

1. Shneiderman B., Plaisant C.: Designing the user interface: Strategies for effective human-computer interaction (4th Ed). Reading, MA: Addison-Wesley, 2003.
2. Norman K. L., Panizzi E.: Levels of Automation and User Participation in Usability Testing, University of Maryland, Laboratory for Automation Psychology and Decision Processes, Technical Report: LAP-2004-01, HCIL-2004-17, 2004.
3. Murphu N.: Principles of User Interface Design. Internet Appliance Design Article, December 2000, http://www.embedded.com/2000/0012/0012ia1.htm.
4. Plaisant C.: The Challenge of Information Visualization Evaluation. In Proc. of Conf. on Advanced Visual Interfaces AVI'04 (2004).
5. Trzaska M., Subieta K.: Active Extensions in a Visual Interface to Databases, Fourteenth International Conference on Information Systems Development (ISD'2005), Kluwer/Plenum Press, 14-17 August, 2005, Karlstad, Sweden.
6. Trzaska M.: Virtual Schemas in Visual Interfaces to Databases, I Krajowa Konferencja Naukowa "Technologie Przetwarzania Danych", pp. 361 - 371, 26-28 September, 2005, Poznan, Poland.

---

[1] Advanced Government Information Service Bus (eGov-Bus, IST 26727 STP) is a project supported by the EC as a part of the Sixth Framework Programme.

7. Carey M.J., Haas L.M., Maganty V., Williams J.H.: PESTO: An Integrated Query/Browser for Object Databases. Proc. VLDB (1996) 203-214
8. Murray N., Goble C., Paton N.: Kaleidoscape: A 3D Environment for Querying ODMG Compliant Databases. In Pro. of Visual Databases 4, L'Aquila, Italy, May 27-29, 1998
9. Fegaras L.: VOODOO: A Visual Object-Oriented Database Language For ODMG OQL. ECOOP Workshop on Object-Oriented Databases 1999, 61-72
10. Cassel K., Risch T.: An Object-Oriented Multi-Mediator Browser. 2nd International Workshop on User Interfaces to Data Intensive Systems, Zürich, Switzerland, May 31 - June 1, 2001
11. Smith M., King P.: The Exploratory Construction Of Database Views. Research Report: BBKCS-02-02, School of Computer Science and Information Systems, Birkbeck College, University of London, 2002
12. Trzaska M., Subieta K.: Usability of Visual Information Retrieval Metaphors for Object-Oriented Databases. Proceedings of the On The Move Federated Conferences and Workshops (DOA, ODBASE, CoopIS, PhD Symposium), Springer Lecture Notes in Computer Science (LNCS 3292), pp. 822-833, October 25-29, 2004, Larnaca, Cyprus.
13. Holyer A.: Methods for Evaluating User Interfaces. Cognitive Research Paper No. 301, School of Cognitive and Computing Sciences, University of Sussex, Brighton, 1993.
14. North C. L., A User Interface for Coordinating Visualizations Based on Relational Schemata: Snap-Together Visualization. PhD Dissertation, Graduate School of the University of Maryland, College Park, 2000.

Author Index

Lecture Notes in Computer Science

For information about Vols. 1–4292

please contact your bookseller or Springer